Technological Innovation and Human Resources 1

Managing Technological Development: Strategic and Human Resources Issues

Technological Innovation and Human Resources

This new De Gruyter series brings together research, critical analysis, and proposals for change in a dynamic and developing field: technological change and its effects on human resources. The series cuts across traditional academic boundaries through its multi-disciplinary perspective. Management researchers and managers, policymakers, legal and labor relations scholars, engineers, psychologists, sociologists, economists, information systems and computer science specialists, and many others share their research findings and insights, giving readers access to a broad field of expertise in a single comprehensive volume.

Technological Innovation and Human Resources examines both the management of technological developments – computers, information systems, telecommunications, artificial intelligence, avionics, and biotechnology – and the changes in management implied by the use of these developments. Special emphases include the focus on human resource concerns. Articles also examine the management of technological change as it links with nationality, ethnicity, type of industry, jobs, skills, organization, and labor market stratification. The series will explore technological, legal, and demographic trends, international and multinational comparisons, as well as theoretical and methodological developments.

Managing Technological Development: Strategic and Human Resources Issues

Edited by
Urs E. Gattiker and Laurie Larwood

Walter de Gruyter · Berlin · New York 1988

Urs E. Gattiker
Asst. Prof. of Organizational Behaviour and Information Systems at the School of Management, The University of Lethbridge in Lethbridge, Alberta, Canada
Laurie Larwood
Dean, School of Business, State University of New York at Albany, Albany, U.S.A.

CIP-Titelaufnahme der Deutschen Bibliothek

Managing technological development: strategic and human resources issues /
ed. by Urs E. Gattiker and Laurie Larwood.
– Berlin ; New York : de Gruyter, 1988
(Technological innovation and human resources ; 1)
ISBN 3-11-011084-9
NE: Gattiker, Urs E. [Hrsg.]; GT

Library of Congress Cataloging–in–Publication Data

Managing technological development : strategic and human resources issues /
edited by Urs E. Gattiker and Laurie Larwood.
p. cm. — (Technological innovation and human resources ; 1)
Bibliography: p.
Includes indexes.
ISBN 0-89925-414-4 (U.S.)
1. Technological innovations—Management. 2. High technology industries—Management. 3. Labor supply—Effect of technological innovations on. 4. Technological innovations—Economic aspects. I. Gattiker, Urs E. II. Larwood, Laurie. III. Series.
HD45.M327 1988
658.5'77—dc19 88–14999
CIP

 – Typesetting: Wagner GmbH, Nördlingen. – Printing: Gerike GmbH, Berlin. – Binding: Dieter Mikolai, Berlin. – Cover Design: Hans-Bernd Lindemann, Berlin. – Printed in Germany.

Preface

This book represents the first volume in the new *Technological Innovation and Human Resources* series. The nine chapters have been grouped into two sections, *Innovation and R&D Management* and *Human Resources and Technological Innovation*. The former concentrates primarily on the "macro" side of technology, examining, for example, the relationship between the acquisition of technology and employment in that industrial sector, or the manner in which high technology development projects are promoted successfully. In contrast, the second section focusses on the "micro" perspective of technological innovation, fitting questions of individual acceptance and use of technology into the equation.

We think you will be pleased with the results. An unusually valuable and creative summary of research cutting across the spectrum of strategic and human resource issues related to the development of technology has been included. Although we cannot claim to be encyclopedic in one book, we think you will find truly provocative work in this first volume with wide-ranging individual, organizational, and social implications.

Contribution selection for this volume. The work represented here was developed in two ways. Much of it is adopted from the top papers presented at a recent meeting of the Academy of Management. After having survived a stiff competitive review process, the papers were then reviewed again by members of our editorial board. A select group of authors was subsequently invited to write about their research for this volume, often requiring a substantially more comprehensive framework than that presented previously. The results were rigorously reviewed up to three times and only the finest articles were retained.

In our initial review process, we found that some subtopics which we thought exceedingly important were not well represented. Consequently, we invited a few leading researchers to develop entirely new contributions on these issues. As with the other group, these papers were closely reviewed to ensure consistently high quality and importance.

The intention of this series is to be international. With that in mind, it might be asked how successful we were in this volume. The answer is, "only partly." As the first in its series, we did not have the time to fully involve most researchers outside North America in this volume. There is an exception, of course, with a participating auther working in the Soviet Union. His work provides an important additional perspective to those originating in the USA or Canada. We hope to expand such representation substantially as the series continues.

We also insisted that each contributor speak to both the theoretical and the practical impact of their research and thinking. We hope you will be as excited as we are over the results. Without exception, our contributors have taken up the challenge of making their research relevant and useful to both academics and decision-makers. From the academic viewpoint, these contributions bring together a broad range of thinking, and several articles make substantial theoretical advances. At the practical level, the studies offer many suggestions for direct application, and some provide surprising insights to the effect that widely-held social beliefs concerning the development and applications of technology may be misguided.

Winter 1987 The Editors

Acknowledgments

We extend our gratitude to the many individuals who provided generous support, editorial help and other critical assistance with this volume: Laura Burris, Cynthia Cunningham, Valorie Hoye, Gail Kembel, Ingeborg Klak, Marlene LaPointe, George Lermer, Bianka Ralle and Sherry Shamroth.

Contents

Urs E. Gattiker and Laurie Larwood
Series Editors' Introduction 1

Section I: Innovation and R&D Management

Urs E. Gattiker and Laurie Larwood
Introduction 7

Lee Tom Perry and Kurt W. Sandholtz
Chapter 1: A "Liberating Form" for Radical Product Innovation 9

Richard N. Osborn, Aubrey Strickstein, and Jon Olson
Chapter 2: Cooperative Multinational R&D Ventures: Interpretation and Negotiation in Emerging Systems 33

Frank L. Edwards and Laurie Larwood
Chapter 3: Strategic Competitive Factors in the Acquisition of Technology: The Case of Major Weapon Systems 55

Michael A. Hitt, R. Duane Ireland, and Igor Y. Goryunov
Chapter 4: The Context of Innovation: Investment in R&D and Firm Performance 73

Barrie Gibbs, Kevin Keen, and Rob Lucas
Chapter 5: Innovation and Human Resource Productivity in Canada: A Comparison of "High" and "Low" Technology Industries 93

Section II: Human Resources and Technological Innovation

Urs E. Gattiker and Laurie Larwood
Introduction 121

Steven W. Floyd
Chapter 6: A Micro Level Model of Information Technology Use by Managers 123

Jo Ann Verdin
Chapter 7: The Impact of Computer Technology on Human Resource Information System Users 143

Urs E. Gattiker
Chapter 8: Computer End-Users: The Impact of Their Beliefs on Subjective Career Success 161

John E. Ettlie
Chapter 9: The First-Line Supervisor and Advanced Manufacturing Technology 187

Conclusion

Urs E. Gattiker
A Brief Summary of Volume I 209

Urs E. Gattiker
Where Do We Go From Here? Directions for Future Research and Managers 213

About the Contributors 219

Author Index 223

Subject Index 229

Series Editors' Introduction

Urs E. Gattiker and Laurie Larwood

This is the inaugural volume of an exciting series that brings together research and critical analysis at the interface of two important fields: technological innovation and human resources. The series is intended to serve as a catalyst for research cutting through established boundaries to yield provocative and important new scholarship, while providing a synthesis across academic disciplines. Those studying new topics often advance knowledge most when they intentionally join related but distinct disciplines. As our field matures, it will itself generate further subtopics, new theories, and research techniques, as well as proposals for practice and policy. This series hopes to contribute to these developments.

Technological Innovation and Human Resources has four major purposes: to spotlight this important and intriguing area of concern, to provide a multidisciplinary academic forum, and to contribute to the development of this field of scholarly endeavour. Last but not least, the series should become a primary source of information for the decision-makers already at work in this field, including managers and engineers.

A series dedicated to such a broad sweep necessarily reflects common experience and common sense. More than half of all workers in North America and Europe are dependent on some form of advanced technology to perform their work tasks. Technology includes but is by no means limited to computers, information systems, telecommunications, computer-aided design and manufacturing, artifical intelligence, avionics and biotechnology.

Some of the research to be examined here concerns issues of technology implementation or enhancement of productivity, which have direct importance both to the success of the organizations and to the richness of work life of the personnel involved. They are of secondary but equal importance to the competitiveness of the society in which the innovations are adopted. This series will also study the management of research and development (R&D) and innovation. How do we encourage innovation and those involved with the process? How do nations encourage it ... and should they?

There are other reasons for focusing on the combination of technology and human resources. Technological innovation in new product design and manufacturing has changed the life cycle of goods in the marketplace. While the overall process of introduction, wide use, and eventual replacement of innovations remains the same, organizations feel pressured by a competitive economic environment to discover and implement technological changes in production and work processes at an accelerating pace. The resulting social changes at the workplace, both in production and office life, are often unplanned and unwanted. Until recently, much social science research in human resource management ignored the influence of technological changes, in favor of detailed examination of an existing and increasingly outdated work environment. We are coming to understand that technological change and innovation affects the entire work system: the individual employee, the department, the enterprise as a whole, and the society in which it operates. Each level must be analyzed to increase our understanding about the relationship between "man and machine."

Multi-disciplinary emphasis. Academia divides its world into disciplines, each with its own assumptions, theories, research methods, values and beliefs. Researchers in the different fields tend to believe that their particular approach to a topic is best. They are often unaware of work done in other disciplines on the same subject. Some feel that research methods applied and accepted in other disciplines are "unscientific" or hard to follow. Many others think, however, that it is important to work with scholars from various disciplines on topics of common interest and concern. Unfortunately, they often find such joint projects difficult to develop or, if successful, it is hard to publish the results in academic journals specific to the individual disciplines.

Technological Innovation and Human Resources provides a much needed alternative to those research projects not specific to any one area. This is not a series about technological innovations, computerization, and human resources, per se. The primary message of most chapters involves social science knowledge; technology is discussed only as a *medium* through which work is shaped and changed. This point is crucial to understanding the series' intended readership and the composition of its editorial review board. It is expected that most of the regular readers of the series will have a primary interest in the interfacing of technological innovation and human resources. Yet the series demands little knowledge of technology as such. An engineer interested in human resource questions related to technology, and a social scientist or personnel manager interested in the same issues will feel equally at home here. Given the focus of the series, more than half of the editorial board members were chosen solely on the basis of their scholarly expertise in various branches of the social and natural sciences. The remainder of the editorial board consists of individuals who have demonstrated expertise specifically in the area of technological innovation and human resources.

The series encompasses multi-disciplinary research by sociologists, mathematicians, engineers, economists, and psychologists, as well as management, educational, policy, legal, computer, human factors/ergonomics researchers, and philosophers. This series was created partly in recognition of the need for a forum to bring together these different facets of the field, and partly in response to the problems faced by researchers and decision-makers in keeping up with the fragmented literature spread throughout many publications and countries.

There are both advantages and disadvantages to a multi-disciplinary series. *Technological Innovation and Human Resources* hopes to present key ideas from one discipline and/or country to those working on the same topic elsewhere. These ideas can then be used, tested, expanded, and evaluated. Nevertheless, articles written from one perspective may seem simplistic or incomplete to those in other disciplines and countries. Equally problematic, the ideas may already seem passé within their source disciplines or nations. We are convinced, however, that the rich creativity of this cross-fertilization of ideas far outweighs any disadvantages.

Developing the field of technology and work. The series' objective is to do more than simply present articles on the theme of technological innovation and human resources. The unifying goal is to provide a forum in which the participants stimulate readers, and over time contribute to the growth of this field. We anticipate that this series will assist in the development of new theoretical perspectives. Some of these results will emerge when previously neglected aspects of computerization and workplace changes are investigated. For example, our hope is to encourage studies of employee experiences with computers and technological innovations for all aspects of work in a variety of organizational settings, industries, and countries. Research which examines new forms of work and organizations as well as changes in skill requirements and labor relations is also highly desirable. Studies in these areas will complement the relatively greater amount of information we have on the many aspects of computerization and its technical features. Such multi-disciplinary and cross-cultural research should aid in advancing our limited knowledge about technological innovation and its effect upon an individual's life and work. Job classification, work structure, gender, political environment, nationality, industry, type of technological innovation, skills, and education are among the variables authors are encouraged to bring into the discussion. In addition, attention should be focused on new forms of work, changes in the labor force and work hours, as well as employment prospects for future generations.

Providing a comprehensive source of information for decision-makers. Today's fast-changing environment is often affected by technological advances. Unfortunately, severe time constraints frequently prevent decision-makers from gaining new insights when a wide variety of sources from different disciplines must be consulted. Moreover, the jargon used may alienate many educated and interested readers who are not "insiders." Narrowly focused publications might provide relevant information to the specialist but offer very little to more general audiences. While individual articles are specific, this series will strive to maintain a broad focus that will be of importance to decision-makers in many areas.

The major types of articles that will be considered for inclusion in *Technological Innovation and Human Resources* are briefly described below. Not all of these categories can be represented in the first volume, but they illuminate the proposed scope of this series.

Reviews of research. This series will publish reviews that integrate and interpret present findings on selected topics. Each contributor will write his or her review somewhat differently, but we hope that all will go well beyond the usual cursory treatment. Research reviews should summarize and synthesize the material on a given subject, exposing to researchers in other disciplines the central thrust of studies done in the area. They should be current in both theory and data and should contain thought-provoking ideas for further research. In addition, reviews should offer decision-makers thorough insights which will facilitate their work in practical contexts.

Case studies. The case study is one technique of inquiry which is not used enough in the field of technological innovation and human resources. This series invites key corporate, institutional and organizational case studies, and histories drawn from both the public and private sectors. Although case studies can seldom offer an exhaustive analysis of a topic, they provide detailed knowledge about a situation in which there has been extensive observation. They can also help to sharpen concepts and inspire further research.

New theories. In their search for sharply focused research, many journals reject unusual pieces, because they may deviate from traditional norms and approaches. We would like this series to become publisher of choice for such leading edge thinking. At the same time, authors of such work must recognize that the most truly creative work bears the additional burden of proving its worth to both editors and readers.

Research series. A researcher may have performed a series of related examinations centered on a topic of concern to multiple disciplines. If space permits, *Technological Innovation and Human Resources* will publish such a series as one unit, provided that the studies are complete and the subject is of widespread interest to a multi-disciplinary audience, as well as practitioners. Alternatively, summaries of such series might be elicited from the researcher when his/her work is concluded.

Cross-cultural and intra-cultural comparisons. The internationalization of business and trade has made it imperative that researchers also internationalize their thinking. Studies of international cultural issues on how individuals and groups function with technological innovation at work will be particularly welcomed. Of course, observing a difference or change in behavior between cultures and countries is one thing, whereas correct interpretation of it is something else. Articles are solicited which study two or more countries, which refer to the interaction between people of different cultures within a single organizational setting as technological innovation is introduced, and which apply a meaningful insight into any differences found.

New research methods and techniques. In their quest for appropriate contributions, many journals discourage the use of innovative techniques and methods on the part of researchers, even if they help increase knowledge about the phenomenon under investigation. Yet the rapid changes caused by technological innovations demand that the narrow paradigm strongly favoring research based on statistical methods and probability theory be expanded to include research which can encompass the diversity of a multi-disciplinary and global perspective.

Technological forecasting and social change. Of major importance to policy-makers are new developments in technology and their relationship to the practice of using technological forecasting as a planning tool. Moreover, the interaction of technology with social, behavioral, and environmental aspects affects strategic planning. The contributions published in this series should facilitate decision-making for managers who are concerned with both technological innovation and human resources. Discussions of implementation techniques and ways of improving the practice of technological forecasting to facilitate strategic planning and policy management are a major objective of the chapters to be included in this series.

Information for Potential Contributors

Inquiries about contributions for any subsequent volume of *Technological Innovation and Human Resources* are encouraged. Future calls for papers will be issued in the relevant publications of various professional associations around the world. Interested authors should submit five copies of their article to Urs E. Gattiker, Technology Research Group, School of Management, The University of Lethbridge, Lethbridge, Alberta, T1K 3M4, CANADA. The format to be used is that of the American Psychological Association. Please prepare your paper to allow for an *anonymous review process*. All chapters must be scholarly, original, and written expressly for *Technological Innovation and Human Resources*. They must be suitable for an academic audience beyond the confines of a single discipline, and should be accessible to managers and other policy and decision-makers. Further editorial guidelines will be provided as necessary.

Section I: Innovation and R&D Management

Urs E. Gattiker and Laurie Larwood

Introduction

This set of five chapters primarily represents a "macro" or systems approach to the effects of technological innovation and the management of R&D efforts. Chapters 1 and 2 ask how we can obtain innovation, and suggest some of the cultural differences influencing its development. Chapter 3 also concerns culture, asking about the strategies firms use when trying to gain a competitive edge in selling advanced weapons systems to the U.S. government. Do we really understand the costs and benefits of technological development? Chapters 4 and 5 focus on this question in different ways, examining the relationships between R&D investment efficiency and firm type, as well as between economic sector and employment growth.

In Chapter 1, *Lee Tom Perry* and *Kurt Sandholtz* discuss which type of organizational culture may foster radical product innovation. This is an extensive case study of a high-tech organization with a reputation for technologically innovative products. The authors draw parallels between a "restricting" and "liberating" organizational culture with an attendant comparison of classical music and jazz, an innovative approach which helps to explain the phenomena under investigation.

Richard Osborn, Aubrey Strickstein, and *Jon Olson* next discuss the complex relationships and negotiation processes involved in establishing a cooperative multinational R&D venture. Their article describes two dominant models of R&D administration and relates them to the problems encountered by multinational joint ventures. A new perspective is presented by considering R&D efforts both as administrative systems and as interpretative orders. The paper outlines the possible problems resulting from divergent expectations upon the effectiveness of a multinational R&D venture, along with strategies to manage cultural differences.

The third chapter, by *Frank Edwards* and *Laurie Larwood*, describes the strategic competitive factors involved in the acquisition of major weapon systems by the U.S. government. This study investigated the perceptions of both contractors and government officials on the use and success of different strategies used to obtain systems development contracts. Additionally, the authors report on the extent to which the contractors' investment effort and attention are related to their obtaining the contract. Their findings indicate how firms intending to obtain major development contracts go about doing so, and what makes them most successful.

Chapter 4 is written by *Michael Hitt, Duane Ireland,* and *Igor Goryunov*, who explore the relationship between a firm's investment in R&D, its growth strategies, diversification, and its performance. Their data reveal that increased diversification leads to increased complexity and difficulty in the efficient management of R&D efforts. Firms which diversify into related fields showed a decrease in R&D efficiency in this study. The paper also deals with the different effects of internal and external growth strategies on corporate performance.

The final chapter in this section was written by *Barrie Gibbs, Kevin Keen* and *Rob Lucas*. It compares the innovation rates of so-called low tech and high tech companies in Canada. The study reveals that increased innovation can lead to growth in employment in both high and low tech firms. The authors also show that innovation by low tech companies may be as effective in job creation and in other critical indicators as is innovation in high tech firms. These results run counter to much common thinking about the importance of the high technology sector to economic development and employment, and indicate the importance of examining these complex relationships carefully before making policy changes favoring high tech firms.

Chapter 1
A "Liberating Form" for Radical Product Innovation

Lee Tom Perry and Kurt W. Sandholtz

Abstract

Radical product innovation is possible when two conditions exist: (1) an "open" product design is maintained; and, (2) organizational members who possess "academic" values are attracted to and retained by the organization. Moreover, there must be an organizational form supportive of the conditions that foster radical innovation. This is a "liberating form."

This view of organizations as potentially "liberating" settings for radical product innovation suggests a number of interesting theoretical and practical implications that challenge traditional assumptions about organizing and managing human resources. Moreover, several important policy implications arise from the disparity between conditions that exist in most large organizations, and conditions that foster radical product innovation.

A "Liberating Form" for Radical Product Innovation

Radical product innovation requires an understanding of the fundamental principles underlying basic technologies. This understanding enables quantum advances to be made within a technology leading to the development of new product families. Radical product innovations either disrupt or make obsolete existing competencies (*Abernathy* and *Clark* 1985). It is appropriate to call radical product innovations "discontinuous" (*Robertson* 1971). They are the "scientific revolutions" that punctuate long periods of evolutionary development (*Kuhn* 1962).

Abundant evidence suggests that radical product innovations originate outside large, privately funded organizations or when originated within are rejected by them (*Abernathy* and *Utterback* 1978, *Goldman* 1985, *Moore* and *Tushman* 1982). This has been explained as: (1) the intended outcome of strategic posturing; and, (2) an unintended consequence of bureaucracy.

Quinn (1979) notes that few managers of large, privately funded organizations assume that they best represent stockholders' interests by engaging in radical product innovation. They instead choose to adopt a strong second-to-market, risk-averse strategy (*Miles* and *Snow* 1978). They let other, usually small entrepreneurial firms absorb the high levels of uncertainty associated with radical product innovation, then either modify or enhance those technological advances that demonstrate proven market potential. This can be a highly profitable strategy for firms with significant financial, marketing, and manufacturing resources.

Other authors take the position that large, privately funded organizations would be unable to stimulate radical product innovation even if they wanted to. They explain that most bureaucratic conditions are hostile to the creativity and change associated with radical innovations. The budgetary systems, accepted time horizons, efficiency criteria, compartmentalization of work, established routines, and the generally risk-averse nature of bureaucracies place significant barriers in the way of major innovation (*Goldman* 1985, *Moore* and *Tushman* 1982).

Whether the intended outcome of strategic posturing or an unintended consequence of bureaucracy, the paucity of radical product innovation in large, privately funded organizations is of major national concern. *Quinn* (1979), for example, warns that if an entire industry ignores radical product innovation that industry is likely to become second rate, and the nation's trade suffers.

The primary purpose of this paper is to describe how firms foster radical product innovation. Our insights have been derived both empirically, conceptually, and most importantly, through comparative analysis of empirical and conceptual findings. Our research began with the establishment of a long-term relationship providing unlimited access to a firm that has been extremely successful at fostering and sustaining radical product innovation. Parallel efforts at conceptualization and reviewing the relevant literature kept us pushing to uncover the deeper meanings hidden in our empirical data.

Conceptually, we concentrated on understanding a variety of forms of creative endeavor. We asked broad, open-ended questions, such as: Are there features common to all radically creative processes? Are there fundamental ways to organize highly creative work? What are these "liberating forms?" We discovered two metaphors – the liberating power of strict poetic form, and the controlled freedom of jazz – that we consider particularly helpful.

Our analysis identified two conditions that seemed basic to creative endeavor and an organizational form that was both supportive and sustaining. We searched the innovation literature to discover support for the importance of these conditions in fostering radical product innovation. We further studied the radically innovative company with which we had a long-term relationship to see if these conditions were indeed crucial to radical product innovation, and also to learn about the nature of an organizational form that supported them.

The structure of our arguments follows the process of discovery described above. We begin with the suggestive metaphors of strict poetic form and jazz improvisation. We discuss the two conditions basic to jazz improvisation and other forms of creative endeavor, then demonstrate how fundamental these conditions are to radical product innovation. Our next step is to describe an organizational form that is supportive of these conditions, and, hence, of radical product innovation. To do this we shift away from what has been a conceptual discussion to a description of the firm with which we have had a long-term research relationship. Finally, we discuss the implications of our findings for (1) managing human resources, (2) large organizations interested in fostering radical product innovation, (3) innovation policy, and (4) future research.

Two Suggestive Metaphors

Clark (1974), in a penetrating essay on life and art, explores the paradox of freedom and form in literature. Some of the most lovely, most energetic poems in the English language are written in one of its most restrictive forms: the sonnet. *Clark* suggests that a poet often finds that conventional poetic forms can anchor, channel, refine, and control creative energy. Borrowing from the world of aeronautics, he draws the following analogy:

> A glider soars, a jet plane flies not just because the wind blows or because the motor develops a half million pounds of thrust. The energy of wind or motor must be exerted on or within significant and controlled form. Imagine all that power being released by a jet engine unattached to anything, like the balloon you blew up as a child and let go – how it hissed and darted and sputtered aimlessly (*Clark* 1974: 39).

Of course, the poet supplies the energy, and mere knowledge of the form never guarantees significant creative achievement. Still, most of the time it is more difficult for poets to create meaningful forms on their own than to create within forms already available and proven (*Clark* 1974).

A metaphor adopted by many organizational scientists likens organizations to symphony orchestras. Although useful for visualizing abstract organizational concepts such as structure, division of labor, and coordination, the symphony metaphor is somewhat less suited for understanding liberating organizational forms. *Mezzrow* and *Wolfe* (1946: 125-126) explain:

> Written music is like handcuffs; and so is the pendulum in white-tie-and-tails up on the conductor's stand. Symphony means slavery [. . .] a creative musician doesn't need compasses or guides or signposts [. . .] the spirit's in him and it will show him the way. One-hundred-men-with-a-fuehrer, a musical battalion hypnotized by a dictator's baton – that's no kind of set up for a man's inspired soul to shake loose and jump out of his instrument in a flood of carefree, truth-speaking, right-from-the-heart music.

Figure 1: Traditional Musical Format

Figure 2: Jazz Format

Figure 3: Invented Melodic Line Based Upon Jazz Format

Mezzrow and *Wolfe* (1946) offer the perspective of jazz musicians. Although they emphasize the freedom of jazz, it is not freedom without form. Because jazz is usually created by a collection of musicians, form is essential; without it, the coordination of effort is impossible. But the jazz form differs from other musical forms in important ways.

A symphony orchestra follows a traditional musical format. Figure 1 is an example of a simple piece – the "Sesame Street" theme song – represented in traditional musical format. This particular example happens to be written in the concert key of F, for alto saxophone and string bass. The music leaves little to the imagination. All notes are specified; the musicians know exactly what, when, and when not to play.

In contrast, Figure 2 is the same "Sesame Street" theme song, represented in jazz format. It could be played by any collection of instruments, in any key. Chords and their duration, rather than notes and rhythms, are specified. The pattern is stable – twelve measures, always played in the same sequence, at an agreed-upon tempo – but open.

This openness gives the jazz musicians license to invent a melodic line (or accompaniment) that "fits" the basic harmonic sequence of the song. Figure 3, a transcription of a sax solo, is an example of what a soloist can do with such deceptively simple harmonic sequences.

Conditions Fostering Radical Product Innovation

How does the jazz ensemble do it? How does it foster improvisation? Despite its seeming effortlessness, meaningful jazz improvisation doesn't just "happen." Good jazz depends upon two conditions being met. First, and most obvious from the discussion above, jazz improvisation requires an *open design* – the musical instructions must be incomplete, thus encouraging members of the jazz ensemble to invent their own melodic lines. Second, the members of the ensemble must be competent jazz musicians, which presupposes the possession of certain values and skills. These are the same values expressed by *Mezzrow* and *Wolfe* (1946) above, namely, that "a creative musician doesn't need compasses or guides or signposts [...] the spirit's in him and will show him the way." These are *professional values supportive of improvisation* as well as the skills to improvise.

We suggest that these two conditions are also fundamental to fostering radical product innovation.

Open Design

There exists a relatively predictable pattern in the amount and type of innovation over a product life cycle (*Abernathy* and *Utterback* 1978, *Moore* and *Tushman* 1982). Radical product innovation occurs early in the product life cycle, up until a dominant design emerges.

At some point in a product's life cycle, firms decide that more can be gained by concentrating on one design and following it down the experience curve than can be gained through continued experimentation. The emergence of a dominant design is an

> "extremely important event [...] in that it fundamentally changes the nature of innovations, the bases of competition, manufacturing processes, and therefore marketing strategy and organizational design" (*Moore* and *Tushman* 1982: 134).

Because the design is no longer "open," product innovation becomes incremental. The focus of the organization shifts to process innovation and the development of bureaucratic efficiencies.

The model depicted in Figure 4 suggests a process by which firms arrive at a dominant design. It is important to note that most authors view the emergence of a dominant design as a positive, if not inevitable, development (*Abernathy* and *Clark* 1985, *Abernathy* and *Utterback* 1978, *Moore* and *Tushman* 1982). They are interested only in eliminating factors that impede product standardization.

Figure 4 implies that a dominant design emerges in one of two ways. One way is to have *evaluation criteria* freely floating. A firm's decision to concentrate on one dominant design is guided by the business opportunities that are created by *design proposals*, whether or not they conform to the original *vision of a product design*. A second way is to have an *explicit* vision of a single-product that is technologically possible. This focuses design proposals on satisfying clear and explicit evaluation criteria, and forms consensus around a dominant design.

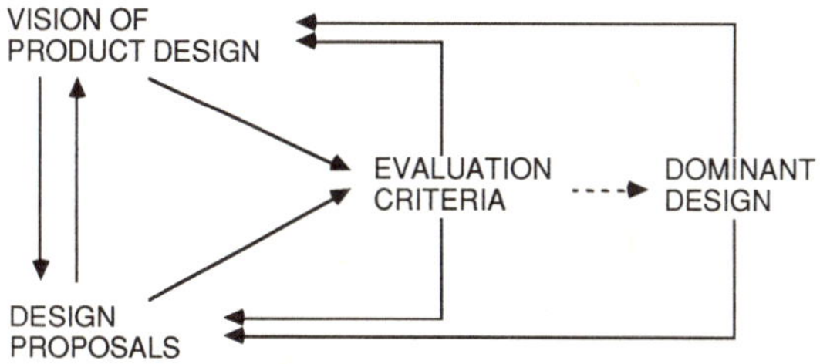

Figure 4: The Process by which a Dominant Design Emerges or an Open Design is Maintained

Only minor adjustments to the *vision of a product design* and *design proposals* (i.e., incremental product innovations) occur, once the dominant design emerges.

Figure 4 also suggests a way to keep a design "open," and by so doing sustain radical product innovation. This occurs when evaluation criteria are fixed, not freely-floating, and reflect a vision (more intuitive than explicit) of an ultimate product design. This is typically a product vision that is not even feasible, given existing technology.

This vision of an ultimate, technologically infeasible product design continually pushes out the frontier of creative endeavor. Like jazz musicians, researchers who are guided by this vision are perpetually dissatisfied with their designs. They are constantly reaching for new and higher levels of creativity. They remain more interested in solving still-unsolved problems than in exploiting the competitive advantages of the problems they have solved. The design remains "open" until their vision is finally realized, which usually takes many years and leads to numerous technical breakthroughs along the way.

Professional Values Supportive of Improvisation

Hlavacek and *Thompson* (1973) adopted the position that major product innovation is a necessary condition for survival in industries with rapidly changing technologies, and attempted to determine what adaptations of the bureaucratic pattern were occurring in 14 firms in the petroleum, automotive, electronics, and chemical industries. They concluded that "new venture teams" offered the most promise as a nonbureaucratic alternative for encouraging radical product innovation. According to *Hlavacek* and *Thompson*, the absence of bureaucratic control mechanisms in new venture teams did not mean that they were "uncontrolled." The usually-high professional nature of the members of new venture teams appeared to provide powerful "professional" control over individuals.

Hlavacek and *Thompson* (1973), however, did not explore the specific nature of the "professional" control that encourages and channels radical product innovation. In this regard, it has been observed that radical product innovations occur in disproportionate numbers in companies and units located near strong science-based universities (*Abernathy* and *Utterback* 1978: 42). Not only is the exchange of technology and people important to the symbiotic relationship between major research universities and companies dedicated to radical product innovation, but also the exchange of professional values and conventions. "Academic" values, conventions, even lifestyles supportive of improvisation are prominent in companies with a commitment to radical product innovation (*Segal Quince* and *Partners* 1985: 71).

What does this mean? Most notably, there is pronounced orientation towards research, design, and development as opposed to marketing, production, and distribution (*Segal Quince* and *Partners* 1985: 35). The nature and extent of this bias, of course, vary from one firm to another, but there is clearly more emphasis given to technological imperatives and less to traditional business concerns. Managers of firms in which this "academic" perspective is securely embedded will even sacrifice opportunities for growth to preserve the nature and style of their enterprises (*Segal Quince* and *Partners* 1985: 71).

These "academic" values and conventions spawn characteristics unique to firms engaged in radical product innovation. Clearly, definitions of technical innovation are quite demanding. Enhancements around existing technologies are not recognized as genuine innovations. To be truly innovative, a product must have a design that is radically different from its predecessors and offers vastly improved performance capability (*Sandholtz* and *Perry* 1986).

Market success is often downplayed as a criterion for success in firms guided by "academic" values and conventions. Rather, they adopt the position that there are no unsuccessful products or projects (*Sandholtz* and *Perry* 1986). This is because even market failures lay a foundation of knowledge and skill that contribute to subsequent technical development. Experimentation with radical alternative designs, then, is a valued activity without concern for where it leads.

"Academic" values and conventions also breed technical arrogance (*Perry* and *Page* 1986). These companies are clearly technology-driven. Their people believe they can develop products and technologies that the market is not even aware it needs. This predictably leads to attitudes that undermine a company's ability to remain close to its market and customers.

A Liberating Form for Radical Product Innovation

An *open design* and supportive *professional values* are necessary but insufficient conditions for fostering jazz improvisation. The jazz ensemble, itself, must also encourage improvisation. There must be an organizational form that is "liberating."

We propose that radical product innovation also requires a "liberating" organizational form. We found such an organizational form while conducting research at Star Electronics.

We have conducted research at Star Electronics over a five year period, with the basic purpose of understanding Star's culture and the innovative processes it supports. The company's fifteen year history is unusual in that gross sales have doubled roughly every two years, yet the company has retained its highly innovative form. Star is still known for its state-of-the-art products, not for its market share. Catering to early adopters with special technical needs, people at

Star take great pride in the company's reputation for technology that is three to five years ahead of markets.

Initially, we were confused by the organizational form that existed at Star. It appeared to violate numerous traditional management principles. Chaos seemed rampant. The management philosophy and style of Star's founder and CEO particularly baffled us. Then we came to realize that the organizational form he had created and nurtured reflected the "academic" values of the university electrical engineering department he had left soon after he launched Star.

We have distilled three unique characteristics of his style to be presented in detail.

Mundane First Assignments

Like many directors of research organizations, Star's CEO believes in recruiting the "best and the brightest" – in Star's case, the top 5% from the nation's premier electrical engineering schools. But unlike other research directors, he often gives these high-powered new hires job assignments that border on the banal. Why? Often, he claims, it is because he doesn't know where they will best fit in. Rather than pre-define an area of contribution, he will assign a mundane task, then wait for the engineer to weary of it.

If engineers are any good, the founder maintains, they will soon return with their own proposals for more challenging assignments. If proposals make sense and relate (however tangentially) to Star's overall mission, engineers are allowed to pursue them. If, on the other hand, engineers continue to perform the mundane work, waiting for someone else to give them another assignment, then they are the "wrong type" for Star. They will not be terminated; instead, they will continue to receive low-level assignments until, frustrated, they leave the company.

Although it may appear somewhat calloused, this CEO's job assignment strategy is anything but a careless default to the "law of the jungle." The process is a deliberate self-selection system, designed to ensure that only those engineers with the desired qualities of "academic" professionalism, initiative, and autonomy are retained. The following anecdote, related by a respected senior engineer, illustrates one instance of the strategy in action:

> A few years ago [Star's founder] came to me and asked if I needed a new recruit on any of my projects. I told him that we could maybe use someone to sweep the floors, but other than that, we had things covered. He said, "I've got a new engineer I want you to take on anyway. Find something for him to do." So I did. I put the kid to work testing circuit boards. Within a few weeks, he was coming to me with suggestions for re-designing the boards. That kid turned out to be one of the brightest, most creative engineers I've ever

> known. From then on, whenever [Star's founder] brought me someone new, I didn't ask questions.

The same philosophy is manifest in the founder's dealings with upper managers. The vice-president of one of Star's two main divisions told us:

> [Star's CEO] will never tell a manager what to do. He figures that if he has to tell you what to do, you're not the right person for the job – and that if you are the right person for the job, you wouldn't do what he told you to do anyway.

In many ways, this approach to management reminded us of Louis Armstrong's often quoted response to the question, "What is jazz?" Implying the futility of abstract definition, Armstrong's reply was simply, "If you have to ask, you'll never know." At Star, this deliberate ambiguity of job assignment has naturally generated increased stress – particularly for "professional managers" brought in from other companies. For others, however, it has provided an opportunity to channel their energy in the directions they deemed most exciting, challenging, and promising.

Unarticulated Organizational Goals

We mention above the importance of an intuitive vision of an ultimate technology in maintaining an "open" design. Closely related to this is the founder's reluctance to articulate explicit company goals. While the founder intuitively knows where he wants Star to go in developing new technologies, he does not clearly communicate it. Part of this is due to the limitations of his intuitive vision – he only knows what he wants when he sees it – but it is more because he considers any sort of "imposed vision" inappropriate for his organization.

A prime example of this reluctance occurred at a recent management retreat. The overriding purpose of the three day "conclave" was to identify a strategic direction for the company. A middle manager describes what happened:

> The main outcomes [of the retreat] were a feeling that top management was listening, and a set of nominal deadlines to produce a strategic plan addressing the question of where we'll be in five years. [...] As I understood it, the deadline for publication of this strategic plan was October 1984. I understood it was supposed to be something that would be distributed to everyone in the company. [Star's founder] was the one who was going to pull the thing together. [...] I know he showed something to the board of directors, but there was no downloading to the rest of the company. It was never published and distributed to employees, not even to middle managers. Any hope of instilling up and down commitment went right out the window. It didn't give everyone a view of "this is where I am, this is how I fit in, etc". [...] Lower level people need to know what's OK to do in the long run. That's never identified.

This manager was frustrated by a perceived lack of direction at Star. He assumed that the founder did not have a clear vision of where the company should go; that he was simply "muddling through." The founder's second-in-command explains, "[Star's founder] is a brilliant strategic thinker, but he never puts it down on paper. It's all in his head".

Fragments of Star's strategic focus are articulated to keep the company from veering too far off course. Often this occurs when a promising project that has been nurtured by a group of engineers is suddenly rejected by the founder because it does not conform to his vision. On these occasions, it is obvious that Star's founder has a clear idea of the markets in which he is interested in competing and the kinds of technologies he wants to develop. But unlike many CEOs, he refuses to articulate his goals unless he has to. And even when he does, it is always in small pieces and to select groups.

Laissez-Faire Management

The previous two characteristics of the founder's style imply an extremely "loose" approach to management. Under this "academic" pattern of management, Star's engineers tend to band together in small, highly autonomous work groups. Beyond granting or withholding resources and sponsoring stringent technical design reviews, upper management makes few attempts to control the groups. The reigning assumption is that the engineers should design and run their own projects.

As a result, thick subcultures develop around these work groups. The groups become the primary socializing units, facilitating learning by apprenticeship. Group members put in long hours and harbor deep feelings about the importance of their work. The environment within these work groups has thus proved an ideal seed-bed for the development of new technology.

At the same time, Star's laissez-faire policy is not without severe dysfunctions. In the words of a middle-manager,

> "The management problems at Star are horrendous: engineers shooting off in wild directions, program managers trying to decide what to pursue and what not to pursue. It's a wild melee".

The consequences of this organizational "melee" include: intense competition between groups for resources; wariness surrounding collaboration and the sharing of ideas (one engineer remarked that "ideas usually get kicked around in close proximity to the people that have them – it's not common for an idea to travel rapidly and uniformly through the company"); vicious political infighting (in some cases, sabotage); and the routine loss of innovative people whose "pet" ideas are not supported. Intergroup tensions damage relationships between research and marketing, research and manufacturing, even manufacturing and marketing. One manager depicted for us the "bad blood" generated by an inter-divisional development effort:

> It was a terrible mess – an absolute morass – between program management and internal product development. People were not respecting each other's backgrounds and time frames. [...] We'd get together in meetings, sort of agree on compromises, then go back to normal procedures after we left the meeting. The manufacturing people would go back to their rules; the engineers would go back to changing their drawings every three hours; and the product management people would go back to missing design opportunities in order to keep the customer happy. It was a cesspool of non-communication.

Allowing these conditions to persist would appear the height of management irresponsibility. And yet, though clearly opposed to the intergroup conflict and professional pettiness catalogued above, Star's founder insists that such dysfunctions are an inevitable by-product of the "academic" values and conventions that ensure radically innovative work. In his view, these negative effects could perhaps be minimized, but only at the cost of disrupting or stifling what has proven to be a remarkably "liberating" organizational form.

Implications for Managers and Researchers

We do not know whether the conditions and organizational form that exist at Star Electronics are consistent across all or even most firms that successfully foster radical product innovation. Therefore, even though significant implications can be derived from our findings, we do not wish to overstate them. They are based upon findings from a single case study, and until they are supported by further empirical research, they remain merely suggestive.

Important implications, however, can be derived from our findings for: (1) managing human resources; (2) large organizations interested in fostering radical product innovation; (3) innovation policy; and (4) future research.

Implications for Managing Human Resources

Despite the apparent chaos, Star Electronics is not "out of control." As with a jazz ensemble, the company's preservation of an "open" design, "academic" values and conventions, and an organizational form supportive of those values and conventions both stimulates and channels creative endeavor. In effect, the culture of the organization's chief participants – rigorously trained electrical engineers – provides the principal motivation, sets the technical parameters, establishes evaluation criteria, and guides the actual process of developing new technology. Like competent jazz musicians, the engineers who elect to remain at Star are encouraged to give free reign to their creativity – to "improvise" – within an accommodating, "liberating" organizational environment.

This view of organizations as potentially "liberating" settings for radical product innovation suggests a number of interesting theoretical and practical implications that extend some of our traditional assumptions about managing human resources.

Selection as an organizational control mechanism. Etzioni (1965: 655) identifies structure, socialization, and selection as the fundamental means of organizational control. It is clear from the discussion above that Star Electronics does not employ structure as an important source of control. Structure offers the kind of explicit form of control that is noticeably absent at Star. Organizational socialization is the process of being indoctrinated, trained, and taught what is important in an organization (*Schein* 1984). Although some system-wide socialization certainly occurs at Star, the presence of thick subcultures also makes socializing processes inconsistent across groups and moderates their effectiveness as control mechanisms. This elevates the importance of selection, especially in terms of grounding the control of the organization in the academic values of its participants. This is consistent with *Etzioni's* (1965: 658) assertion that "utilitarian" organizations delegate socialization to other organizations, such as professional schools, and prefer careful selection of socialized persons over the efforts needed for their socialization.

Most highly innovative research organizations recognize the importance of selection, as evidenced by a variety of recruiting techniques. Some organizations hire by referral only, or do the bulk of their recruiting at select graduate programs (*Segal Quince* and *Partners* 1985: 45-46). Others put candidates through a rigorous marathon of interviews, show them work in progress to see if their eyes light up, and probe about the kinds of sacrifices the candidate would be willing to make to see a project through. Only those who pass every test of passion, skill, and "fit" are awarded jobs.

The key difference between such practices and Star's system is the element of *self*-selection. The assumption is that new recruits receive mundane first assignments and, if they are the "right" type for Star, they will *self*-select more interesting and meaningful assignments. As *Van Maanen* and *Barley* (1984) point out, individuals with an occupational (or professional) perspective, who would be the "right" type for Star, assess their careers more in terms of improving their technical skills and gaining control over the work they do than in terms of hierarchical advancement or recognition. Self-selection at Star, then, grants to individuals the freedom to exercise their "academic" initiative.

Redefining leadership. In an organization whose control system relies on the values of a larger "academic" culture, leaders need to be exemplars of these professional values. This role of "representing" the profession's values operates at two levels: *internally*, to reinforce the organization's "liberating form"; and *externally*, to attract the "best and the brightest" of the profession to the organization (*Dalton* and *Thompson* 1986).

At Star Electronics, for example, most of the formal (and all of the informal) leaders are referred to as "Technical Gurus." Many of them maintain high "academic" profiles, presenting "breakthrough" papers at conferences and earning patents for new technologies they have developed. Star's founder is himself a member of the Computer Science Hall of Fame and an acknowledged leader in the field of computer graphics.

The technical preeminence of Star's CEO is particularly interesting in light of *Vaill's* (1982) research into "high-performing systems":

> Motivation in high-performing systems [...] has "peculiarity." In leaders, this peculiarity can be seen in the way they have integrated their innate energies and ambitions with the system's needs and opportunities. *Selznick's* "institutional embodiment of purpose," therefore, is not a one-way phenomenon. In high-performing systems there is a two-way embodiment in Feeling of leader-in-system and system-in-leader. When a person becomes "Mr. [...] (whatever the activity or industry is)" or "the First Lady of [...] (the activity or industry)," the two-way embodiment is complete (p. 32).

Similarities between *Vaill's* high-performing systems and our notion of "liberating form" are, in our view, not surprising. Determining the elements of intersection between the two concepts is certainly an area that merits further research.

Reframing organizational loyalty. Loyalty in firms dedicated to radical product innovation becomes a function of the degree to which the organization allows "academic" values and conventions to be expressed. The theoretical congruence with *Vaill's* (1982: 28) research is again difficult to ignore: one of the keys to identifying high-performing systems is that such systems "are perceived to fulfill at a high level the ideals for the culture within which they exist – that is, they have 'nobility'."

Our long-term research relationship with Star Electronics has allowed us an unusual vantage point for monitoring this perceived "nobility" and the organizational loyalty it engenders. Over the last five years, as Star has grown (and, in some areas, bureaucratized), we have witnessed a measurable increase in feelings of dissatisfaction and betrayal among engineers. Whereas the common self-description used to be, "We are the Cadillac of the industry," many engineers have begun to claim that "Star is not as innovative as it used to be," or that "technical expertise is no longer the most important job qualification around here." In the words of one of the "technical gurus":

> Prior to 1981, there were no job titles in the company. We were all "members of the technical staff." Then they started bringing in more business types, and calling people manager of this or that. It really became noticeable in the summer of '81. It was like the company had switched from an emphasis on technical excellence to a "road construction" mindset – a philosophy of,

> "These are trainable jobs. To get the software done, we'll hire more bodies and rent some bulldozers." They stopped hiring the best and the brightest.

This respected engineer's opinions is by no means universally shared. Throughout the company, however, we have recently registered a sense of yearning for "the good old days," when engineers wielded more control (and, thus, enjoyed more freedom). For some, the creation of a formal R&D group signaled that "the rest of the company" no longer had the mandate to "push the technology," but were instead to focus on applications and upgrades. This perception has led a number of engineers to idealize the R&D group as the last bastion of "the way things used to be."

The limitations of professional managers. The "liberating form" perspective also exposes the inevitable limitations of "professional managers" who have not been socialized into (and therefore do not understand) the "academic" values and conventions of those they are supposed to manage. When the organization's culture is derived from non-managerial occupational values, organizational members may tend to view MBA-trained managers (with their traditional "organizational" values) as deviant.

We observed an example of this at Star Electronics. The introduction of managers with backgrounds in areas other than electrical engineering provoked resentment and reinforced elitism among many of the engineers. In the words of a frustrated software designer:

> It seems like management has been taken over by marketing people. I think it's a mistake. Development works best with competent, exciting technical managers directing small, cross-functional groups of engineers. [...] If the manager isn't a competent technical person, when the people under him get stuck, they're stuck. [...] When you have reasonably intelligent people reporting to less intelligent people, you have problems.

On the other side of the coin, all of the non-technical managers we talked to attested to the difficulty of earning respect in a system that did not reward (nor understand) their contributions. At Star, these "business" people typically do not last long in their positions – unless they can somehow gain the respect and support of the engineers. One of the most successful of the non-technical managers described for us an interesting (and apparently effective) method of compensating for his technical ignorance. Rather than trust his own judgement in technical matters, he would bring in high-powered technical consultants ("gurus") for design reviews, and base his decisions on their assessment. In addition, he established a close relationship with Star's founder. Thus, the engineers in his division began to realize that although he did not have the technical expertise that they possessed, he was at least willing to play by their rules and act as their advocate with the CEO.

Implications for Large Organizations

When we compare Star Electronics to many large organizations operating in the private sector, we see the differences between a set of conditions that fosters radical product innovation and a set of conditions that does not. For us, the contrast is striking. What is also clear is that conditions that support some forms of innovation undermine others. Those conditions that lead to incremental product innovation in many large organizations are incompatible with those conditions that foster radical innovation at Star Electronics.

This observed incompatibility is the basis for an alternative explanation for why radical product innovation does not occur in large organizations. It is not that large organizations cannot foster radical innovation; rather, they will not tolerate the conditions necessary to do it.

Based upon what we observed at Star Electronics, we have identified four actions that most large organizations could, but would be reluctant to take to stimulate radical product innovation. Perhaps these actions imply a more strategic orientation than exists at Star. Moreover, it may be impossible to recreate in another organization a set of conditions that arose naturally. Nevertheless, if it were our intention to create conditions that foster radical product innovation, these are some of the actions we would take.

Adopt a first-to-market, low-volume production strategy, catering to early adopters, then exit markets to avoid head-to-head competition with large, second-to-market competitors. This is a strategic posture consistent with maintaining an "open" design. In order to maintain an "open" design, engineers must concentrate on high-performance, high-margin, low-volume products, and be willing to abandon these products before they gain broad market acceptance. For example, Star's engineers take pride in the fact that they are typically five years ahead of markets for their products. They take little or no pride in their ability to compete with other companies, once comparable products have been developed.

Managers of large organizations would be critical or intolerant of this action. New products must off-set huge development and overhead costs. Managers would perceive moving out of markets before products gain broad market acceptance as losing interest in products just as they become profitable. Their logic would be to exploit each product over an entire product life-cycle. This is done by concentrating on enhancements and alternative uses for the product, as well as process innovations that lower the cost of production. They would wait to introduce "next generation" (i.e., radical) products until the energy behind "present generation" products is exhausted.

Let the vision of an ultimate technology, not a product/market strategy, drive innovation. The development of radical product innovations is comprised of many unpredictable twists and turns. A vision of an ultimate technology unobtrusively guides the work of "academic" engineers. They neither feel controlled nor constrained by the vision. They "sign up" to be members of the team, meaning they concur with the vision. They are left to establish their own short-term direction while guided by the longer-term vision. This offers a wide latitude of acceptable activity. The potential for radical innovation exists because "academic" engineers feel the freedom needed to explore and experiment with new ideas.

Quinn (1980) observes that control in large organizations does not reside in one person or group. It is something that is negotiated, shared and arises out of a complex web of relationships and goals. It is important to realize that radical product innovations are not free of political implications. Often, radically new products developed by one group will infringe on markets in which other groups have significant vested interest. The purpose of product/market strategies is to protect vested interests and define politically neutral avenues along which innovation can proceed. Large organizations prefer innovation to follow product/market strategies because the political ramifications are predictable and usually can be held in check. Managers of large organizations do not like to be surprised, and, therefore, are quite intolerant of the haphazardness associated with innovation guided by a vision of an ultimate product design.

Moreover, managers would be even more reluctant to rely on a vision of an ultimate technology when: (1) they do not wholly understand the technical side of the vision; (2) the "academic" visionary does not have a proven track record; (3) the vision cannot be articulated clearly; and, (4) there is turbulence in the environment to which a product/market strategy can more appropriately and quickly adjust.

"Hive off" teams of "academic" engineers, granting them full autonomy to develop radical product innovations. *Peters* and *Waterman* (1982: 215) use the example of 3M to argue that the key to innovation is not growing bigger divisions, but to spawn, or "hive off" new ones. *Peters* and *Waterman*, however, do not address the more extreme requirements of radical product innovation. They are more interested in what *Cooper* (1984) calls "high-success rate" and "high-performance" product innovation.

"Hiving off" teams of "academic" engineers will not foster radical product innovation, unless the autonomy granted to these teams far exceeds the prescriptions of *Peters* and *Waterman*. This is because managers of large organizations can neither understand nor appreciate the "academic" subcultures that will develop in these teams to foster radical innovation. These teams will operate on an entirely different set of assumptions. Their work habits will appear to be undisciplined and lack direction. Managers will perceive the inefficient behavior of teams of "academic" engineers and quite naturally believe they are wasting the organization's investment. "Academic" engineers will define success in terms of technical criteria, rather than potential market success. *Quinn* (1980) has observed that the use of technical evaluation criteria is particularly difficult for executives to accept. Because they do not understand technical criteria, they perceive a much higher level of risk than do technical people. Finally, "academic" engineers will "beat up" on each other, compete with each other, even sabotage each others' work. This will raise management's ire. Instinctively, managers will want to go in and clean up the mess.

Moreover, there are no promises that benefits will ever come to large organizations from these teams of "academic" engineers until breakthroughs actually occur. If and when a breakthrough does occur, the team will probably not know how to develop it further, and, even if it did, would not want to. What this means is that these teams will never become profit centers. They will always consume, not produce assets.

Perhaps the most disturbing consequence of "hiving off" teams of "academic" engineers is management's total lack of control once the venture is launched. If at any time managers were to interfere, in an attempt to channel the activities of what appeared to them to be a floundering team, they would certainly destroy all hope of radical innovation. Given this, it is not surprising that managers of large organizations would be reluctant to follow this course of action.

Encourage a sink-or-swim approach to career development for members of "academic" teams. This action is consistent with the granting of mundane first assignments to new hires. The strategy relies on a process of natural selection. The "right type" of engineers self-select themselves when they either define or join challenging projects that capture their interest. The "wrong type" of engineers eliminate themselves when they choose to leave the organization after receiving one-too-many mundane assignments.

A prominent feature of Star's sink-or-swim approach to career development is the way it accelerates the turnover of young engineers. The consequences are similar to those resulting from the rigorous tenure systems of major research universities. Obsolescence is seldom a problem because there are few mid-career engineers. The majority of the engineers are young, and have been with the organization for less than five years. These young engineers have brought with them state-of-the-art research experience, new perspectives, and fresh ideas. They are often anxious to prove themselves, and, therefore, extremely competitive with each other. They know that if they make a name for themselves early, that many exciting opportunities will be offered to them – if not by Star, then by other companies.

The work of these energetic, commited young engineers is guided by a few senior engineers, who like Star's founder set the tone and direction of the organization. These are people who have already made a name for themselves. They provide role models for the young engineers, and offer a vision to them that they probably will be unable to complete in their lifetimes. The two generations of engineers are linked together by a common vision, and the work continues.

The managers of large organizations are likely to see all the liabilities and none of the assets that accrue from this sink-or-swim approach to career development. They will be suspicious of the apparent arbitrariness of the system; the technical elitism it implies; the widespread under-utilization of unappreciated kinds of engineering talent; and the high rates of turnover. Thus, they are more likely to encourage, even impose, their own systems of programmed advancement, mentoring, and sponsorship.

Implications for Innovation Policy

We proposed above what large organizations would need to do to foster radical innovation. This, of course, is a dubious agenda that managers of most large organizations will find impractical and choose not to follow. What this implies is that we expect little radical product innovation to occur in large organizations.

This perspective raises several important questions related to innovation policy. We need to ask:

(1) If radical product innovation cannot occur in large organizations, then where can it occur?

(2) How do we fund radical product innovation?

(3) How do we accomplish the transfer of radical product innovations?

Alternative sources of radical product innovation. Given that it is difficult to foster radical product innovation in large organizations, the most obvious policy implication is to direct efforts at accelerating the formation of small, entrepreneurial start-ups. Of course, only a few of the many entrepreneurial start-ups will offer "liberating forms" for radically innovative work, especially if they need to look like Star Electronics. Before we arrive at this conclusion, however, we need to ask if only an organizational form like Star's can be liberating? Or are there alternative organizational forms that are also liberating? Moreover, we may want to learn more about how to hold back the natural (perhaps inevitable) forces that destroy "liberating forms" as organizations grow. We already know it is extremely difficult to sustain a firm's capacity to do radically innovative work.

Our findings also imply that universities, where "academic" values and conventions are firmly rooted, may offer a natural environment for fostering radical product innovation. This suggests that increased support for university-based research and development may be vital for the U.S. to continue to be radically innovative and, thus, remain industrially competitive. Another approach may be to provide tax incentives for industry to support university-based innovative work.

Joint university/corporate research relationships may be helpful in stimulating radical product innovation, provided corporations become more like universities, rather than vice versa. Our concern here is that in expanding or extending the stream of funding on which these joint relationships depend, university research centers must not be pressured to adopt corporate values and practices. If they were to do so, our findings suggest that they would lose their capacity to be radically innovative.

Finally, it may be advantageous for large, privately funded organizations to invest in radical product innovation by joining high-tech research consortia (e.g., Semiconductor Research Corporation; Microelectronics & Computer Technology Corporation), and then focus on transfering, developing, marketing, and manufacturing radical product innovations generated by the consortia. Of course, this also is assuming that these research consortia are able to create and maintain “liberating” organizational forms.

Funding radical product innovation. Our findings suggest that the *kind* of funding given in support of radical product innovation is at least as important as the *amount* of funding. For example, *Abernathy* and *Utterback* (1978: 45) note that vast amounts of money are spent supporting large, integrated corporate R&D departments, but these units develop mostly “technologically-active middle range” innovations, not radical innovations. On the other hand, *Quinn* (1979: 21–22) observes that developers of radical product innovations often have started working out of their homes to keep overhead costs to a minimum. They have relied more upon “sweat capital” than dollars for materials and equipment.

Star’s founder suggested to us that the best kind of funding for radically innovative work comes in the form traditionally given by foundations. He explained that this was because the funding was tied to a general concept, not a specific research proposal. The funding was also secure and long-term in nature. Funding from foundations allowed him to retain control of his company while pursuing his long-term vision of an ultimate product technology.

In other high-tech firms we have studied, an early indicator that the firm was moving away from radical product innovation occurred in the project review process. The emphasis shifted from reviewing the technical merits of a project to review of a business plan. Once this shift in emphasis occurred, the number of radical product proposals began to drop sharply. Radical product proposals could not be formatted as business plans without appearing to be of unacceptably high risk.

The problem here is that the kinds of research proposals demanded by many funding agencies resemble business plans, especially in terms of the detail required. Radical product innovation, by its very nature, involves the exploration of “virgin” territory. It cannot be mapped out and planned, except in a very general sense. If funding agencies wish to encourage radical innovation, they should, like foundations, support general concepts and people more than detailed proposals.

The transfer of radical product innovations. Although our findings tell us little about how to transfer technology, they do suggest that technology transfer is of crucial importance. Radically innovative engineers do not want to follow their new technologies very much beyond the design stage. Most of them want to return to basic research. This means that unless mechanisms are in place to facilitate the effective "handing-off" of new technologies, many radically innovative ideas are lost.

When we propose that we direct increased emphasis and funding to high-tech start-ups, university-based research centers, joint university/corporate arrangements, and research consortia, all with "academic" cultures, we also must consider how we accomplish the transfer of technology out of these environments. Of course, a few entrepreneurs will be able to exploit the market potential of their radically innovative ideas; a few university researchers will take their new technologies out of the university setting and start new companies; corporations can also buy understanding of new technologies by hiring new Ph.D.s – but these are all informal and somewhat arbitrary mechanisms for accomplishing the transfer of technology. In many cases, they may not work as well in the future as they have in the past. Clearly, we need to rethink the technology transfer problem. This is an area that requires more systematic and serious study as well as deliberate, imaginative experimentation.

Implications for Future Research

The model we have presented of a "liberating form" for radical product innovation is only a starting point. Clearly, there are many questions left unanswered, many implications left unexplored. Our hope is that the model and its implications will promote further research and debate.

The final thought we are left with pertains to how difficult it is to create, maintain, and benefit from a "liberating form" for radical product innovation. Only through a sustained commitment to learning about the unique characteristics of a "liberating form" can we hope to understand how to manage more effectively for radical product innovation.

References

Abernathy, W. J. and *Clark, K. B.* (1985): Innovation: Mapping the winds of creative destruction, *Research Policy*, 14: 3–22.

Abernathy, W. J. and *Utterback, J. M.* (1978): Patterns of industrial innovation, *Technology Review*, 80: 40–47.

Clark, M. J. (1974): Liberating form, *Brigham Young University Studies*, 1974, 15: 29–40.

Cooper, R. G. (1984): How new product strategies impact on performance. *Journal of Productive Innovation Management*, 1, 5–18.

Dalton, G. W. and *Thompson, P. H.* (1986): *Novations: Strategies for career management*, Glenview, IL: Scott, Foresman.

Etzioni, A. (1965): Organziational control structure, in: *J. March* (Ed.), *Handbook of organizations*, Chicago: Rand-McNally.

Goldman, J. E. (1985): Innovation in large firms, in: *R. S. Rosenbloom* (Ed.), *Research on technological innovation, management and policy*, (Vol. 2), London: JAI Press.

Hlavacek, J. D. and *Thompson, V. A.* (1973): Bureaucracy and new product innovation, *Academy of Management Journal*, 16: 361–372.

Kuhn, T. S. (1962): *The structure of scientific revolutions*, Chicago: University of Chicago Press.

Mezzrow, M. and *Wolfe, B.* (1946): *Really the blues*, New York: Random House.

Miles, R. E. and *Snow, C. C.* (1978): *Organizational strategy, structure and process*, New York: McGraw-Hill.

Moore, W. L. and *Tushman, M. L.* (1982): Managing innovation over the product life cycle, in: *M. L. Tushman* and *W. L. Moore* (Eds.), *Readings in the management of innovation*, Boston: Pitman.

Perry, L. T. and *Page, R. A.* (1986, August): *Why radical product innovation does not occur in large organizations: The intolerance factor*, Paper presented at the annual meetings of the Academy of Management, Chicago.

Peters, T. J. and *Waterman, R. H.*, Jr. (1982): *In search of excellence: Lessons from America's best-run companies*, New York: Harper & Row.

Quinn, J. B. (1979): Technological innovation, entrepreneurship and strategy, *Sloan Management Review*, 20: 19–30.

Quinn, J. B. (1980): *Strategies for change: Logical incrementalism*, Homewood, IL: Richard D. Irwin.

Robertson, T. S. (1971): *Innovative behavior and communication*, New York: Holt, Rinehart & Winston.

Sandholtz, K. W. and *Perry, L. T.* (1986, March): *Professional values, cultures, and conventions: A “liberating form” for radical product innovation*, Paper presented at the annual meetings of the Western Academy of Management, Reno, Nevada.

Schein, E. H. (1984): Organizational socialization and the profession of management, in: *D. A. Kolb, I. M. Rubin,* and *J. M. McIntyre* (Eds.), *Organizational psychology: A book of readings* (4th ed.), Englewood Cliffs, NJ: Prentice-Hall.

Segal Quince and *Partners* (1985): *The Cambridge phenomenon: The growth of high technology industry in a university town*, London: Brand Brothers.

Vaill, P. (1982): The purposing of high-performing systems, *Organizational Dynamics*, Autumn, 11, 23–39.

Van Maanen, J. and *Barley, S. R.* (1984): Occupational communities: Culture and control in organizations, in: *B. Staw* and *L. L. Cummings* (Eds.), *Research in organizational behavior* (Vol. 6), Greenwood, CT: JAI Press.

Chapter 2
Cooperative Multinational R&D Ventures: Interpretation and Negotiation in Emerging Systems*

Richard N. Osborn, Aubrey Strickstein, and Jon Olson

Abstract

The view of R&D multinational joint ventures as interpretative negotiated orders is outlined to reconcile apparently inconsistent recommendations and casual mechanisms underlying the administration of R&D. This view calls for examining quite different kinds of logical systems to understand comparatively new mechanisms for administrating R&D. The analysis suggests that as organizations move into cooperative arrangements for R&D with foreign corporations they will face complex managerial problems. Faced with vague goals and the inability to transfer existing management into the new systems, managers will invent new interpretations of R&D administration. These new managerial solutions will resolve apparent inconsistencies in current recommendations for effective R&D management.

Introduction and Purpose

The literature concerning the administration of research and development is not only extensive but replete with controversy concerning the administrative mechanisms that are most important for assuring R&D success. Those following an economic perspective, for instance, often emphasize entrepreneurial incentives supported by an efficient, tightly controlled administrative system (e.g., *Williamson* 1975). In contrast, those following a professional model often emphasize the importance of exchange of information among specialists supported by a flexible, decentralized, and consultative administrative system

* Preparation of this manuscript was supported by a grant from the National Science Foundation, (No. SES-8504397), Richard N. Osborn, Jon Olson, and Mitsuyo Hanada co-principal investigators. The authors want to express their appreciation to anonymous reviewers for their assistance. Correspondence concerning this manuscript should be sent to: Richard N. Osborn, Department of Management, Wayne State University, Detroit, MI 48202.

(*Hage* and *Aiken* 1970). In the economic perspective of R&D, the causal mechanisms include clarity of purpose, control of interaction, and minimization of transaction costs. In the professional analysis, the causal mechanisms often center on professional exchange of information, the climate needed to foster experimentation, and the freedom given to professionals to pursue their ideas unencumbered by excessive bureaucratic intrusion.

Here an analysis of emerging multinational cooperative ventures for research and development suggests that the conflict between the two perspectives is more apparent than real. Specifically, emerging R&D units are likely to employ somewhat different logical systems in attempts to capture the beneficial effects of both economic rationality and professionalism.

The argument is presented as follows. Two dominant models of R&D administration are outlined with an emphasis on the causal mechanisms used to justify recommendations for successful R&D management. A brief comparison of these models with the problems faced by emerging multinational joint ventures for R&D suggests that a whole series of governance and interpretative factors have been ignored in much of the previous research. The view of cooperative R&D joint ventures as negotiated interpretative orders is outlined and then used to reconcile the apparent inconsistencies in the economic and professional perspectives of R&D administration.

The Professional Perspective of R&D

What is labeled here as the professional perspective of R&D administration has its roots in the socio-technical literature (e.g., *Trist* and *Bamforth* 1951), sociological inquiry into the administrative structure of R&D (e.g., *Hage* and *Aiken* 1970), and contingency organizational theory (e.g., *Burns* and *Stalker* 1961). While this literature is too extensive to systematically review here (see *Tornatzky* et al. 1983), such analyses suggest that the probability of effective R&D administration is enhanced with what *Burns* and *Stalker* (1961) originally labeled an "organic" structure. When coupled with consultative leadership (e.g., *Bass* 1981) and coordinative mechanisms stressing professional exchange rather than bureaucratic form, the more flexible structure facilitates lateral exchanges among professionals and reinforces technically based communication.

Role of Uncertainty

These recommendations for appropriate administration of R&D often flow from a combined analysis of external conditions, task characteristics, and individual motivation. Specifically, the environment of the R&D system is characterized as technically uncertain, calling for contact with external parties and a broad range of responses to (a) capitalize on emerging opportunities and (b) blunt new demands and constraints (see *Osborn* and *Hunt* 1974). While exchanges may be orchestrated by individuals with specialized roles (e.g., boundary spanners, see *Aldrich* and *Herker* 1977), technical specialists also need the information being generated external to their units to devise new ways of responding to a changing environment. Effective information exchange for R&D is characterized as informal, verbal, and centered on technical concerns rather than planned and administratively directed and centered on mundane corporate concerns such as budgets, implementation, and profitability (e.g., *Keller* and *Holland* 1978, *Utterback*, 1971).

Role of Complex, Ambiguous Tasks

A second series of causal mechanisms concerns the tasks of technical specialists. The tasks are presumed (a) to be unprogrammed (e.g., *Perrow* 1970); (b) to call for highly specialized knowledge (e.g., *Hage* and *Aiken* 1970); and (c) to require that individuals specify both means and ends (e.g., *Thompson* 1967). With such complex tasks, bureaucratic controls merely frustrate task accomplishment. Facilitative mechanisms promoting exchanges among specialists are expected to yield technical synergy. Finally, it may not be necessary to use elaborate organizational controls on professionals, since their training incorporates professional values consistent with insuring technical accomplishment. Thus, bureaucratic controls may not only be redundant but counter to professional norms of autonomy.

In sum, administrative systems should facilitate technical interactions. The administrative apparatus should be designed to serve the needs of professionals. It should assist them in responding to technical uncertainties and not constrain professionals as they naturally seek solutions to vexing technical issues.

The Economic Perspective of R&D Administration

In stark contrast to the professional model of R&D management is one that can be labeled the economic perspective (e.g., *Williamson* 1975). This perspective stems from work in economics, particularly transaction cost economics, and engineering. As one might suspect, economic and technological mechanisms are stressed. Economic analyses begin with the notion that the organization is a substitute for individual entrepreneurs – it is simply a more efficient method of managing transactions than other forms of contracting (e.g., *Williamson* 1975).

Role of Organization Structures

Thus one begins with the premise that the organization is a legal fiction containing contractual relationships, "which is characterized by the existence of divisible residual claims on the assets and cash flows ..." (*Jenson* and *Meckling* 1976: 311). The organization must develop administrative mechanisms to insure that differences in the knowledge among parties and unforeseen changes do not lead to opportunism. It is presumed that contracting parties may, some writers assume many will, take advantage of knowledge asymmetries or unforeseen conditions (cf. *Williamson*, 1975, 1981). Unless the contracting parties can resolve possible opportunism, they will be unlikely to develop the specialization necessary to complete the R&D project. Specialized training, equipment, or facilities will only be developed where trust is assured.

Further, the organization must develop an efficient series of mechanisms to replace complex negotiations over the terms of the contract. As *Williamson* (1975) suggests, decision by administrative fiat is far more efficient than negotiation. Further, organizations need to assure focused attention by individuals and insure that their efforts are not directed toward self-serving ends. For instance, incentive pay schemes such as contingency bonuses based on the market performance of R&D products and processes are consistent with the recommendation to keep the focus of attention on organizational goals.

While descriptions of the administrative arrangements are not elaborated in most economic analyses, descriptions allude to what *Burns* and *Stalker* (1961) would call a mechanistic structure. Formal roles, titles, and positions are important as are written rules, policies, and procedures. Careful documentation and elaborate planning are often stressed. Above all, position in the administrative hierarchy determines the types of information and privileges given to professionals and managers. Such a structure facilitates efficient flows of communication centered on the assumed goals of the R&D unit – research and development that ultimately yields profitable products and/or services or more efficient processes.

The underlying causal mechanisms are quite clear. Financial gain motivates individuals, while the administrative structures (a) provide efficient substitutes for the market and (b) prevent opportunism.

Role of the Environment

For economists the environment is an important driving force for R&D when it provides ample opportunities for economic gains. Some recommend extending monopoly rights and reducing imperfections (e.g., clarifying contractual expectations to reduce transaction costs). Others recommend reducing the cost of research, development, and deployment. For instance, government funding of

R&D and cooperative government/industry partnerships have been proposed (see *Tornatzky* et al. 1983).

Theoretically, markets are always considered viable potential alternatives to organized R&D. But the economic perspective places considerably more emphasis on different forms of cooperation among principals and agents. Thus, joint ventures, cooperative agreements, and venturing agreements are all viewed as administrative arrangements that might substitute for short-term market transactions. For research and development these arrangements may be superior because they are mechanisms for reducing transaction costs. They may survive and be effective because they can provide unity of direction, focused communication (due to their smaller size), and entrepreneurial incentive. As *Raelin* (1985) suggests such focused efforts are becoming quite common in large scale organizations.

Engineering Perspectives and the Importance of Technology

Interestingly, the administrative recommendations emanating from an economic perspective are often consistent with two themes stemming from the engineering literature. First, there is a series of studies concerning improvements in the production process via the implementation and development of new technology (cf. *Buffa* 1977). The underlying causal mechanisms stress energy reduction improvement of quality by minimizing failures and increasing tolerances and/or applying basic research (see *Osborn, Olson,* and *Hanada* 1985). Clear definitions of the problem are to be matched with advanced engineering and scientific techniques. The effort is tightly administrated to derive maximum benefits for minimal effort. Not surprisingly, such studies emphasize individual engineering and technical skill combined with the development of sophisticated administrative techniques (e.g., operations research models).

A second line of engineering research describes the success of large-scale R&D projects. While mainly descriptive, this literature emphasizes a combination of technical, individual, political, and economic themes with the premise that large-scale efforts are required to make significant technical innovations. Successful R&D efforts are often characterized as (1) "technically sweet" (sophisticated in concept but comparatively simple in implementation), (2) derived from the scientific or engineering genius of one or a few researcher/administrators, (3) supported politically by one or a handful of insightful champions, and (4) efficiently administered. Such accounts often include a reification of engineering professionalism coupled with elaborate descriptions of the supporting administrative system. Interestingly, these accounts often reveal a network of cooperation among several independent specialized organizations. Within these organizations the administrative systems are often characterized as highly specialized, centrally directed, and replete with technically driven controls (e.g., elaborate scheduling models and hierarchical clearances of technical reports) (*Osborn* and *Connor* 1984).

Summary of Model Differences

Table 1: Effective R&D Administration: Proposed Differences Between Professional and Economic Models of R&D Administration

	Professional	Economic
Environmental Assumptions	Favorable and volatile with modest dependence on others	Competitive and somewhat volatile with high dependence on others
Technology	Rapid change most important force	Modest change rates and only a secondary factor
Structure	Flexible structure, decentralized decision making	Rigid structure, centralized decision making
Leadership	Consultative to add info flow	Directive to reduce opportunism
Coordination Mechanism	Professional exchange and dialogue	Rules, policies, procedures
Control patterns	Informal, verbal emphasized	Formal written emphasized (e. g., incentive pay schemes)
Role of Management	Facilitate information exchange among specialists	Minimize transaction costs to enhance R&D profits and minimize opportunism

Table 1 summarizes key differences in what we have called the professional and economic models of R&D administration focusing on the importance of environmental conditions and technical affairs. The bottom line for most managers is that the models appear to lead to inconsistent methods of R&D administration. The primary focus in the professional model is facilitating information exchange among technical specialists. Administrative arrangements should allow specialists to respond to technical uncertainties and free them from unnecessary constraints. Formal controls are often considered redundant. In the economic perspective the environment is the driving force, and market mechanisms are considered viable alternatives. Administrative arrangements should be crafted to minimize transaction costs and insure against opportunism. Thus, the recommended administrative arrangements are often centralized formal structures replete with numerous bureaucratic controls to guard against opportunism. Interesting studies of large-scale engineering projects suggest that networks of organizations may be combined to provide overall flexibility, even though each is rigidly structured internally.

Both models share some implicit assumptions that are quickly becoming outmoded with movements toward global markets, political changes in the attitude toward cooperative relationships among competitors, and advances in technology itself. The next section outlines some of the changes.

The Evolution of Multinational R&D

In many industries there has been a movement toward a global market. While autos and electronics are often cited as the primary examples, global markets are beginning to emerge where combinations of the following are present: (a) skilled craftmanship in production; (b) extensive developmental engineering; (c) heavy fixed investments; and/or (d) substantial rapid innovation. In these conditions cooperative relationships can help firms share risks, access wider markets and extend the life of existing products, and adjust products to specific market niches.

Many new products and processes are emerging at the intersection of once separate research specialties (e.g., ceramic engines and biotechnology). Since most firms have specialized expertise in only one field, they often need the specialized experience of others. Cooperative ventures are a natural outgrowth of past specialization. With more firms in Europe and Asia competing technically with U.S.-based firms, the cooperative ventures are often multinational.

Beyond the economic and technical trends, several reports suggest that national political trends are favoring multinational cooperative arrangements (e.g., *Osborn, Olson,* and *Hanada* 1985). Developing nations (e.g., Korea) desire advanced technology and want to develop the capability to conduct their own R&D. In contrast, developed nations are concerned with the "boomerang" effect where once less developed nations capture key markets in developed economies (e.g., the U.S. experience with Japan). Apparently, formal cooperative agreements are a mechanism for reconciling these interests, since corporations from both nations can maintain some control over the technology and its subsequent deployment.

In sum, conditions are becoming much more favorable for the development of cooperative arrangements among organizations with different national home offices. Throughout the 1970's one found the development of the multinational industrial organization with its home offices in one developed country and a host of foreign involvements. Conditions are now beginning to favor the development of cooperative relationships focusing on research and development.

From the perspective of the corporate sponsors cooperative ventures appear to have a number of advantages. Cooperative R&D ventures allow sponsoring corporations to pool risk, increase their potential for market access, counter political pressures against outsiders, and learn from other corporations. For instance, *Hladik's* (1985) analysis suggests that these administrative innovations may help large corporations avoid decadence.

Despite the advantages listed above, multinational cooperative ventures are quite risky affairs. There are the standard risks of R&D. Beyond these risks, the effort is multicultural, multi-organizational, often multi-industry, and newly developed. Each of the chafacteristics poses a number of sticky administrative problems.

Cultural Factors

Unlike national cooperative arrangements which are made up of members from similar national ties and backgrounds, multinational arrangements may be plagued with a vast amount of societal, cultural, and organizational differences which make these arrangements difficult to manage. In particular, when one stops to consider the amount of research that has been conducted on the conflicts internal to single organizations (e.g., *Blake* and *Mouton* 1984), adding the multinational level to the organizational analysis only compounds potential anomalies. Therefore, one would predict that numerous mechanisms would be established to monitor potential differences between the firms engaged in multinational cooperative agreements. Following *Williamson's* logic, resolving coordination and trust questions will be key issues to confront.

Cross-cultural research (cf. *Hofstede* 1980) has shown that different cultures can be classified according to their unique and shared social values. Using a four dimensional model of cultural values, *Hofstede* (1984) reports cultural variation among over 53 countries and regions on the following dimensions: power distance, dealing with uncertainty, individualism-collectivism, and views on masculinity. For instance, *Hofstede* reports that the United States was ranked the most individualistic, above average on tolerance for uncertainty, moderate on tolerance for unequal power distribution, and strong on masculine values. In contrast, Japan was classified as low on individualism, large on power distance, strong on uncertainty avoidance, with a masculine value structure. In a similar vein, *Ronan* and *Shenkar* (1985) clustered countries according to various attitudinal dimensions of work goals, values, needs, and job attitudes. They concluded that countries could, in fact, be clustered on the basis of language, religion, and geography proximity. They found strong support for the Anglo, Germanic, Nordic, Latin European, and Latin American clusters. In addition, they noted that clusters describing the Far East and Arab countries need further refinement. Further, some nations do not readily cluster with others (e.g., Israel and Japan).

In summary, cultural variations may impede the success of multinational corporate arrangements. It has been shown that organizations tend to reflect natural culture values and national political agendas (*Hofstede* 1984). Further, it has been suggested that the corporate characters of U.S.-based firms are quite different from that of other nations such as Japanese-based firms (*Bartels* 1982, *Howard* and *Termoto* 1982).

Organizational Differences

The difference in national cultures is often compounded by the unique character of the organizations who wish to cooperate. Corporations may have unique corporate cultures, a history of successful operations, and a particular strategy. The unique corporate cultures, operating practices, and strategies of sponsoring firms can easily clash. For instance, reports suggest that some Japanese firms prefer to establish a long-term relationship without detailed specification of exactly what is to be accomplished (i.e., detailed contract) and be bounded by a specific budget (*Ballon* 1979, *Wright* 1979). For some Japanese firms agreements are to be cooperative partnerships. In contrast, some U.S. firms prefer detailed legal arrangements covering as many contingencies as can be foreseen. Not only do styles clash, but negotiating the arrangement becomes problematical.

The foundation of a cooperative R&D venture does not necessarily signal that all conflicts have been resolved and that all parties clearly understand the purpose, role, costs, and possible contributions of the joint venture. As participants in the development of a joint venture have commented, "It wasn't easy [...] there were tough negotiations [...] we thought we had the deal sealed tight. But everybody read it differently [...] If the parties reach a deadlock, I'm not sure how it can be resolved" (*Wysocki* 1986).

Inside the joint venture, executives may face difficulties selecting employees, training recruits, encouraging multicultural teams of researchers, establishing new administrative routines, and simultaneously developing a whole host of appropriate rules, policies, and procedures. In different terms, managers confront the liability of newness. To compound this liability, the ventures are often comparatively large efforts reflecting the critical mass needed for R&D (see *Osborn* et al. 1985).

The Paradox of Problems and Opportunities – R&D as an Interpretative Negotiated Order

So far, the reader has been presented with two apparently conflicting approaches to the administration of R&D. Yet, instead of attempting to resolve these comparatively simple questions of how to administer R&D, we are suggesting that firms embark upon multinational cooperative ventures for R&D. In doing so, these firms confront problems of overcoming cultural barriers (both national and corporate) and the liability of newness.

We propose that attempts to deal with wide variations in national cultures, corporate strategies/cultures, and the liability of newness will provide a point of resolution concerning how to administer R&D. The key comes in understanding cooperative R&D ventures as something more than either a mechanism for professionals to interact or a contracting vehicle designed to minimize opportunism.

Cooperative R&D Ventures as Interpretative Negotiated Orders

According to *Weick* and *Daft* (1983), organizations can be viewed as interpretation systems. In particular, they consider the key role of top administrators as being one of making sense (i.e., meaning) out of the events which occur in the organization's environment. The top administrator receives information via various scanning techniques (e.g., data collection) and instills it with "meaning" to provide the organizations with a direction and purpose. "Organizational interpretation is formally defined as the process of translating events and developing shared understandings and conceptual schemes among members of upper management" (*Daft* and *Weick* 1984: 286).

Several implications for organizational researchers are noted by *Daft* and *Weick* using an interpretive logic. First, to enhance the probability of survival, organizations should develop mechanisms to interpret the environment and to provide meaning and direction for employees. Therefore, future research should emphasize the type of administrative mechanisms organizations use to capture meaning from their environment and internal operations. Second, scanning and sense making activities are considered critical in understanding organizational behavior. Almost every aspect of the administrative system is subject to some form of interpretation by the members of the system.

A major problem with the *Daft* and *Weick* (1984) view of interpretation systems is that executives' views are not explicitly bounded. While some executives may offer interpretations that appear totally inconsistent with other accounts, one suspects their integrity, honesty, and judgment may be questioned. Thus, we argue, executive interpretations are bounded and negotiated. They are likely to be bounded by widely-held views of reality and negotiated with peers.

In multinational cooperative ventures interpretations are likely to be both more restricted and less precise. The apparent additional restrictions and loss of precision allow for the development of multiple interpretations and the promulgation of value based symbols, myths, and rituals as a basis for planning and control. The next sections examine each of these paradoxical properties of multinational cooperative ventures.

Multiple Interpretations of Procedural and Substantive Due Process

Several scholars suggest that as firms move across diverse national cultures they encounter information that is difficult if not impossible to interpret within their own cultural experience (e.g., *Ronan* and *Shenkar* 1985). The root of these potential differences lies in definitions of substantive and procedural due process. Substantive due process deals with the concept of equity. Within the Western European tradition there are but subtle differences in this notion across Europe and North America. As *Ouchi* (1980) shows, shared definitions of equity backed by strong social institutions can lie at the heart of effective transactions. A shared notion of equity can substitute for stricter measures to circumvent problems of opportunism.

Procedural due process centers on the standard routines to assure that each party has been given a fair opportunity. In the Western European tradition there are considerable differences in the procedures constituting procedural due process, but invariably these involve bureaucratic routines often adjudicated by a third party. Such is not the case in many Asian countries. When definitions of substantive and procedural due process clash, misunderstandings can often occur. Limits to interpretation can be encountered.

This apparent dichotomy in procedural and substantive due process between U.S. and Japanese firms has been articulated in two different models of how organizations come into existence (cf. *Maitland, Bryson,* and *Van de Ven* 1985 for a review of these models). Both *Williamson* (1975) and *Ouchi* (1980) discuss the role of cooperation in their theory of organizations. Where the two theorists differ is in their assumptions about human motivation. In *Williamson's* view, parties will cooperate based on expected calculations of their expected returns. In contrast, *Ouchi* feels that parties will cooperate when tasks and rewards are perceived to be equitable. Both theorists also disagree on the manner in which parties will operate in attaining their perspective goals. According to *Ouchi*, participants will try to establish mechanisms whereby they can measure their perspective contributions to the project. Further, due to internal conflicts and problems with measuring individual contributions to performance (i.e., performance ambiguity), *Ouchi* notes that parties will behave according to "norms of reciprocity." They will develop and abide by internal norms of the organization.

In contrast, *Williamson* feels that due to their basic opportunistic nature, managers will establish mechanisms to guard against such behavior. Transaction costs according to his model are incurred with trying to guard against opportunism. Therefore, *Williamson* suggests that formal contracts and internal bargaining relationships will evolve.

To make the argument less abstract, it is useful to develop an extended example. For instance, Japanese managers may enter the cooperative arrangement when they (a) trust the other party and, (b) understand their potential partner and expect to maintain the relationship for a long time. *Tung* (1984) has shown that in order to negotiate with the Japanese, substantive due process is evolved from informal dialogue and discussion among respective partners. In contrast, U.S. managers may enter the cooperative agreement for totally different reasons altogether. They may see the arrangement as a short-term opportunity to attain the technology they require for a particular project, with little concern for future arrangements with the Japanese partner. The U.S. manager may be more concerned about establishing and specifying the boundaries of the arrangement in a formal and legal contract (*Tung* 1984). The rules and procedures of how the two partners will operate under all possible conditions will be drafted up in a binding contract. Further, it is common for U.S. managers to use intermediaries such as lawyers to establish and process the cooperative agreements. This is in sharp contrast with Japanese managers who typically use informal face-to-face dialogue to establish the trust required for these arrangements. If disputes arise that cannot be resolved by dialogue, U.S. managers may consider using the court system to resolve disputes. Japanese managers are more likely to use the good auspices of a third party member who has an active interest in both corporations. These apparent substantive and procedural due process differences between the U.S. and Japan may account for the early contractual problems noted in joint venture start-ups (*Osborn* et al. 1985).

We propose that *Williamson* (1975) and *Ouchi* (1980) have suggested potentially complimentary mechanisms for developing and maintaining substantive and procedural due process. At the same time, differences in the definition of substantive and procedural due process may cause problems, working through different interpretations and definitions may provide two complementary mechanisms for defining equity and for resolving disputes. Perhaps this is best seen by examining organizational goals, since clarity of purpose is often seen as a prerequisite for developing an effective R&D effort.

Goals in an Interpretative System

Goals become bounded and emergent when the venture is seen as an interpretative system. One may assume that the heads of these R&D ventures are in continual interaction and negotiation with sponsors and employees over both the allocation of resources and the legitimacy of their units (cf. *Pfeffer* and *Salancik* 1978, *Thompson* 1967). As relatively new systems, the ventures have the opportunity to develop their own unique symbols, myths, meanings, and legitimacy; particularly if differences in the administrative systems of the hosts block a direct transfer of one partner's corporate culture (*Trice* and *Beyer* 1984).

Following from other cross-cultural studies (e.g., *Hofstede* 1984), it is quite likely individuals from different cultures and organizational systems will have different interpretations of the venture, its mission, and their roles in the new endeavor. In different terms, top venture managers face questions of interpretation.

Interpretation of Goals

Most analyses of R&D units presume that the goals of the system are clear. The view of these units as negotiated interpretative systems modifies this somewhat.

Specific goals for R&D units may include (1) invention (the creation of new products, processes, and/or services); (2) development (the refinement of new products, processes, or services); and (3) technology transfer (adoption and/or modification of existing products, processes, or services) (cf. *Williamson* 1975). The interpretative perspective suggests that even the formal specification of these goals leaves considerable room for individual interpretation. If one also assumes that we lack clear criteria for judging R&D success, remain uncertain concerning how it should be administered, and realize it may be years before financial returns materialize, the zone for interpretation becomes even broader.

Several studies suggest that specific, detailed specification of goals may not precede action but emerges from experience (*Starbuck* 1983, *March* and *Olson* 1976). These attributions of success are bounded by the ability of executives to link actions to interests of sponsors and employees. Within these bounds activity deemed desirable by sponsors or employees is reinforced and becomes a goal for the future. Paraphrasing *Weick* (1979) and *March* and *Olson* (1976), organizations want what they have already started to obtain.

In the case of multinational cooperative ventures, apparent progress in any area that can be linked to the broad areas of invention, development, or technology transfer may emerge as future goals. In the early stages of the venture managers are expected to identify some desirable activities from the vast stream of actions occuring within, and use them to justify proclamations of success (*Daft* and *Weick* 1984).

In sum, the view of formalized goals as a basis for actions and decisions may be useful for units with a long history. But for new systems, the issue of goals becomes merely another arena for interpretation and negotiation.

Implications of Interpreted Goals

Without clear goals to justify actions or guide resource allocations, there is likely to be more emphasis placed on less tangible "philosophical" aspects of management in the multinational cooperative R&D ventures. The negotiation of interpretations through the management of symbols, metaphors, language, and ritual may become particularly critical in the founding stages of these ventures (e.g., *Pfeffer* and *Salancik* 1978). While these apparent instruments of interpretation may merely help promote the illusion of control by higher management in new multinational units, such may not be the case. The symbolic respects of administrating these innovative systems may also become an important aspect of how power is developed and allocated.

Governance in Multinational Joint R&D Ventures

Organizational governance is concerned with the development and allocation of power (*Formbrun* 1984, *Scott* et al. 1981). A first aspect of governance derives from questions of substantive due process or the values and meanings that should be given priority (cf. *Scott* et al. 1981). In common language, "what really counts around here."

The most readily observed aspect of governance concerns questions of procedural due process. That is, how should decisions for allocating organizational resources be made? What causal mechanisms (justifications) and procedures should be adopted? The two models outlined earlier in this chapter represent administrative outcomes of two different approaches to the procedural due process.

The Interpretive Logic of Governance

A number of writers have suggested that the logic of substantive governance is quite different from that of procedural governance (*Scott* et al. 1981). For procedural governance the logic may rest on rational deductions of positive and negative consequences. It is the economic logic of costs and benefits, the engineering logic of technical trade-offs, and the sociological logic of structural functionalism (e.g., *Parsons* 1956). However, the logic of substantive governance may be represented by the phrase "multiple positives."

Some forty years ago, *Lynd* (1939) observed that cultural values are both positive and conflicting. He matched apparently conflicting positive statements and categorized them into dominant and contradictory categories. Recently, renewed debates over the legitimacy of corporations has suggested that values often prized in organizations (e.g., hierarchy, collective action, and practical results) are contradictory to those prized in U.S. society (e.g., democracy, individualism, academic achievement).[1] In Lynn's terms, our society's contradictory values may be organizationally dominant ones.

In the logic of substantive due process there is no particular problem with subscribing to contradictory values. Individuals dissemble. Elimination of apparent contradiction may be labeled ideological extremism. Much may be the same for some organizations. Those without what *Deal* and *Kennedy* (1982) call "strong cultures" may follow the logic of substantive due process in terms of how executives generally manage meaning, symbols, myths, and rituals. For instance, in the management of meaning problems may be labeled opportunities. Mistakes may be called learning experiences or experiments.

Role of Symbols, Myths, and Rituals

Several recent articles document the management of symbols, myths, and rituals (*Schein* 1985, *Trice* and *Beyer* 1984). While the management of meaning, symbols, myths, and rituals may not appear to have a direct impact on the allocation of resources, many argue it has powerful effects on the individual's perception of what is valued in the organization and thus provides a mechanism for structuring.

In emerging ventures where sponsors and participants may not share common cultures or experiences, the logic of substantive due process may be a way of capturing the benefits of apparently inconsistent causal mechanism. Specifically, numerous competing values may be reinforced by higher management in their interpretations of events, actions, or conditions (assignment of meaning). The identification of issues may also recognize potential multiple positive outcomes. For instance, an individual may be rewarded for his contribution to team development, recognizing the importance of both individualism and collective action. The venture may be heralded as both an example of cultural cooperation and fighting competitive spirit. Where the environment is uncertain, tasks are ambiguous, and informal patterns of influence are not clear, the substantive side of governance may be used to help provide structure, target attention, and control action. Executives may also use substantive governance in terms of altering meanings, symbols, rituals, and myths to provide professional incentives. In addition, executives may emphasize the freedom of profes-

1 *Argyris* (1965), among others, notes a similar conflict between organizational predispositions and individual needs.

sionals to pursue new technical initiatives and set some of their own research priorities.

Resolving Model Inconsistencies

A brief examination of the logics of substantive due process suggests that it may be used to tap apparently inconsistent causal mechanisms for R&D success. Managers may interpret the environment of the system so that it is compatible with the structural form actually adopted for the venture. If, for instance, the structure follows the dictate of the economic perspective, executives may interpret the environment as clear, predictable, and supportive. Evidence of other interpretations may be rejected. Although subordinates may lack information from external sources, the motivational restrictions this structure is said to have on professionals may not emerge if the substantive aspects of governance are appropriately manipulated. Specifically, the emergent symbols, rituals, and attributions for apparent progress can stress professionalism. Both collective scientific collaboration and individual initiative can be embodied in the "war stories" of success. The executives can promote professional exchange even though the structure is designed to carefully manage transactions, the hierarchy of direct communication, and insure that effort is directed to those targets selected by management itself.

Conversely, the structure selected by higher management may follow the recommendations of those subscribing to the professional model. Managers would still like to capture the benefits of clarity of purpose, control of interaction, and minimal transaction costs. Such may be possible with selective interpretation. Clarity of purpose may be partially achieved interpretatively by selecting specific activities as exemplary of the type of progress expected in the venture. Control of interaction may be partially captured by providing differential status to selected professionals. Minimizing transaction costs may be partially achieved by helping professionals develop a common specialized technical language, resulting in enhanced coordination and integration of member contributions.

In sum, the substantive, interpretative aspects of governance may be employed to *compensate* for the structural and administrative forms derived from the procedural aspects of governance. The logic of substantive due process allows the development of compensatory interpretations. It also suggests that the apparent inconsistencies among the symbols, myths, and rituals of the venture and its structure can continue unchallenged for some time. "What really goes on around here," then varies depending upon whether one is characterizing the symbolic system or the "real" structure and administration.

While such dualism may be present in many R&D efforts, it is likely to be most apparent in the emerging multinational cooperative R&D ventures. Over time, the interplay among the interpretative aspects of governance and the structure may yield new patterns – new ways of administering R&D. These new ways may allow R&D efforts to capitalize on apparently conflicting causal mechanisms for R&D success.

Multiple interpretations are more likely maintainable in units where representatives of different cultures interact. The concept of equitable treatment must meet standards developed from both host cultures and may be supported by procedures derived from both.

In sum, the development of multinational cooperative efforts for R&D provides an opportunity to resolve the apparent paradox of R&D administration. The ventures may be seen as interpreted negotiated orders. Multiple interpretations of such key factors as goals may characterize these complex entities. With multiple cultural roots the manager may need to negotiate several acceptable definitions of substantive and procedural due process. One negotiated multiple interpretation can be made and used to complement decisions concerning the professional or economic models. If an economic model is selected, the substantive due process of the venture can exude professionalism. If the professional model is selected, the definition and use of symbols, myths, and rituals to support the definition can stress economic interpretations.

Implications

This paper argues that the conflict between the professional and economic modes of R&D administration may be more apparent than real if one considers R&D efforts as both administrative systems and interpretative negotiated orders.

The need for this new perspective is evident in a description of multinational joint ventures for research and development. This was most clearly evidenced in an analysis of their goals. The revised review of the R&D units introduces the logic of substantive due process as a critical aspect of the development and allocation of power. The politics of governance does not call for logical consistency in the traditional sense but permits multiple interpretations.

Managerial Implications

To the extent that the interpretative aspects of governance can *compensate* for structural and administrative dysfunctions, several implications can be drawn.

Clarify Expectations

One suggestion from our review centers on the role of identifying both the explicit and implicit expectations of getting involved in cooperative agreements. Before engaging in such arrangements, it would be wise to identify the specific expectations each partner has about the joint venture. For example, one partner may be looking at the arrangement as a means for gaining market accessibility, whereas the other partner may be primarily concerned about learning the latest technology and using the knowledge to enhance their corporate technological power base in the industry.

To the extent that divergent expectations occur in practice, prospective venture partners may be opening themselves up for disappointments. The critical point to remember is that each partner has their own unique agendas and expectations about cooperating. The beliefs may not be compatible with one another. Further, due to the fact that many joint ventures involve partners from different countries with unique value structures (*Hofstede* 1980), it is critical to understand that each partner may place different importance on the value of cooperation. Recall from *Hofstede's* research, for instance, that the U.S. and Japanese have diametrically opposing degrees of importance placed on the individualism and collectivism dimensions.

Develop Cultural Scenarios

A second implication from this line of research is that managers can create scenarios of how someone will react, given their knowledge about the nature, background, and culture of the firms involved. From an open systems perspective, it has been suggested that organizational analysts should consider both the internal and external factors which may impact organizational behavior. Managers in multinational cooperative ventures are confronted with potential differences due to firm specific cultural variations, norms, and customs of respective firms, corporations may vary considerably with respect to "what is really important around here." Therefore, to the extent that managers have an understanding of these important internal factors, they will be prepared to handle such conflicts when they arise.

Several authors have suggested that to better understand cultural and corporate variations (*Schein* 1985, *Tung* 1984), in-house resource people may be valuable informants. Those with prior experience in the culture of the potential partner can be invaluable. Firms should identify such people and use them in transmitting this information across respective lines of venture participants.

Manage Meaning

Another implication from this research is that multinational managers should make themselves aware of the concepts of procedural and substantive due process in organizational behavior processes. Since managers in these cooperative ventures may come from countries with radically different cultures, considerable time should be taken to understand the importance and meaning placed on key concepts. For instance, it has been argued that the pre-venture procedures are dramatically different in Japan than they are in the United States (*Tung* 1984). *Tung* has articulated the different cultural variations to consider in negotiating with the Japanese and considers the role of procedural and substantive due process as a critical construct to pay attention to. Managers should construct meanings with their counterparts and devote resources to bridging cultural gaps and not merely assume the parties are communicating.

Devote Resources to Bridge Cultural Gaps

Following up on these previous suggestions, organizations can develop mechanisms within the joint venture to bridge cultural and value gaps. For example, organizations can construct management training programs which help respective managers with these cultural and value differences. Role simulations could be constructed allowing managers to play the role of different partners to get a better understanding of the views each manager has about the expectations and belief systems each may hold. From a personnel selection perspective, organizations could also structure selection programs to identify and recruit managers who have an understanding of respective cultures and multinational arrangements.

Research Implications

For the theorists looking at these emerging forms of cooperation for research and development, several questions remain. The theory developed earlier suggests that organizational governance may be used to reconcile apparent differences between competing models of R&D administration. The key was multiple interpretation developed by the executives in charge of the venture. The negotiation among these individuals and their respective host organizations may stress a form of cultural dualism. As in many other areas, the theorist must now examine the extent to which executives have developed dual interpretation systems. One suspects many forms will be found, but which forms emerge and remain as viable forms to promote effectiveness is an open question. Researchers need to return to the R&D community and study how astute executives are adjusting their expectations of cooperative R&D ventures, developing scenarios to understand potential clashes between partners, and managing meanings. We have proposed that with the additional interpretation potentials in cooperative multinational R&D ventures managers will start to resolve apparent differences in theoretical prescriptions for managing R&D. We hope these new ventures will be investigated.

References

Aldrich, H. E. and *Herker, D.* (1977): Boundary spanning roles and organization structure, *Academy of Management Review*, 2: 217–230.

Argyris, C. (1965): *Organization and innovation*, Homewood, IL: R. D. Irwin.

Ballon, R. J. (1979): A lesson from Japan: contract, control, and authority, *Journal of Comparative Business*, 8: 27–35.

Bartels, R. (1982): National Culture – Business Relations: United States and Japan Contrasted, *Management International Review*, 22: 4–13.

Bass, B. (1981): *Stogdill's handbook of leadership*, New York: The Free Press.

Blake, R. R. and *Mouton, J. S.* (1984): *Solving costly organizational conflicts*, San Francisco: Jossey-Bass.

Buffa, E. S. (1977): *Modern Production Management*, New York: John Wiley and Sons, 5th ed.

Burns, T. and *Stalker, G. M.* (1961): *The management of innovation*, London: Tavistock Publications.

Daft, R. L. and *Weick, K. E.* (1984): Toward a model of organizations as interpretation systems, *Academy of Management Review*, 9: 284–295.

Deal, T. and *Kennedy, A.* (1982): *Corporate cultures: The rites and rituals of corporate life*, Reading, MA: Addison-Wesley.

Formbrun, C. J. (1984): Structures of organizational governance, *Human Relations*, 37 (3): 201–223.

Hage, J. and *Aiken, M.* (1970): *Social change in complex organizations*, New York: Random House.

Hladik, K. J. (1985): *International joint ventures: An economic analysis of U.S. foreign business partnerships*, Lexington, MA: Lexington Books.

Hofstede, G. (1980): *Culture's consequences: International differences in work-related values*, Beverly Hills, CA and London: Sage.

Hofstede, G. (1984): The cultural relativity of the quality of life concept, *Academy of Management Review*, 9: 389–398.

Howard, N. and *Termoto, Y.* (1982): The really important differences between Japanese and western management, *Management International Review*, 22: 19–30.

Jenson, R. and *Meckling, R.* (1976): Residual claims and discounted flows, *Journal of Economnic Issues*, 30: 301–326.

Keller, R. T. and *Holland, W. E.* (1978): *Technical information flows and innovation processes*, Washington, D. C.: National Science Foundation (PRA-7618441).

Lynd, R. (1939): *Knowledge for what: The place of science in American culture*, Princeton, NJ: University Press.

Maitland, I., *Bryson, J.*, and *Van de Ven, A.* (1985): Sociologists, economists, and opportunism, *Academy of Management Review*, 10: 59–65.

March, J. G. and Olson, J. P. (1976): *Ambiguity and choice in organizations*, Bergen: Universitetsforlaget.

Osborn, R. N. and *Connor, P.* (1984): Leadership in emerging technologies, in: *J. G. Hunt* (ed.), *Leaders and managers: International perspectives on managerial behavior and leadership*, London: Pergamon Press.

Osborn, R. and *Hunt, J.* (1974): The environment and organizational effectiveness, *Administrative Science Quarterly*, 19: 231–246.

Osborn, R. N., Olson, J., and *Hanada, M.* (1985): *Analyzing U.S./Japanese joint research and development units*, Washington, D.C.: National Science Foundation.

Ouchi, W. G. (1980): Markets, bureaucracies and clans, *Administrative Science Quarterly*, 25: 129–142.

Parsons, T. (1956): Suggestions for a sociological approach to the theory of organizations, *Administrative Science Quarterly*, 1: 63–85.

Perrow, C. (1970): *Organizational analysis: A sociological approach*, Belmont, CA: Wadsworth.

Pfeffer, J. and *Salancik, G. R.* (1978): *The external control of organizations: A resource dependence perspective*, New York: Harper and Row.

Raelin, J. A. (1985): The basis for the professional's resistance to managerial control, *Human Resource Management*, 24: 147–175.

Ronan, S. and *Shenkar, O.* (1985): Clustering countries on attitudinal dimensions: A review and synthesis, *Academy of Management Review*, 10: 435–454.

Schein, E. H. (1985): *Organizational culture and leadership*, San Francisco: Jossey-Bass.

Scott, W., Mitchell, T. R., and *Perry, N. S.* (1981): Organizational governance, in: *P. Nystrom* and *W. Starbuck* (eds.), *Handbook of organizational design*, (Vol. 2), New York: Oxford University Press.

Starbuck, W. H. (1983): Organizations as action generators, *American Sociological Review*, 98: 91–103.

Thompson, J. D. (1967): *Organizations in action*, New York: McGraw Hill.

Tornatzky, L. G., Eveland, J. D., Boyland, M. G., Hetzner, W. A., Johnson, E. L., Roitman, T. and *Schneider, J.* (1983): *The process of technological innovation: Reviewing the literature*, Washington, D.C.: National Science Foundation.

Trice, H. M. and *Beyer, J. M.* (1984): Studying organizational cultures through rites and ceremonial, *Academy of Management Review*, 9: 653–669.

Trist, E. L. and *Bamforth, K. W.* (1951): Some social and psychological consequences of the Longwall method of goal-getting, *Human Relations*, 4: 3–38.

Tung, R. L. (1984): *Key to Japan's economic strength: Human power*, Lexington, MA: Lexington Books.

Utterback, J. M. (1971): The process of innovation: A study of the origin and development of ideas for new scientific instruments, *IEEE Transactions on Engineering Management*, 18: 124–131.

Weick, K. E. (1979): *The social psychology of organizing*, (2nd ed.), Reading, MA: Addison-Wesley Publishing Co.

Weick, K. E. and *Daft, R. L.* (1983): The effectiveness of interpretation systems, in: *K. S. Cameron* and *D. A. Whetten* (eds.), *Organizational effectiveness: A comparison of multiple models*, New York: Academic Press, 71–93.

Williamson, O. E. (1975): *Markets and hierarchies*, New York: Free Press.

Williamson, O. E. (1981): The economics of organization: The transaction cost approach, *American Journal of Sociology*, 87: 548–577.

Wright, R. W. (1979): Joint venture problems in Japan, *Columbia Journal of World Business*, 14: 25–31.

Wysocki, B. (1986): Weak in technology, South Korea seeks help from overseas, *Wall Street Journal*, January 7: 1.

Chapter 3
Strategic Competitive Factors in the Acquisition of Technology: The Case of Major Weapon Systems*

Frank L. Edwards and Laurie Larwood

Abstract

Diffusion of innovation models cannot readily predict the strategic management manipulations involved in the acquisition of major high technology weapon systems. This study examined the application of MacMillan's (1978) two-factor model of strategic manipulation to the technology acquisition process. Project managers, representing government, firms winning industry defense contracts, and firms losing them, assessed items for their importance to development contracts for major weapon systems. Results support the model, suggesting that political-organizational manipulations beyond those publicly acknowledged are important in obtaining a contract. Further data examine winner-loser differences in perception and the relationship of firm commitment to obtaining contracts. Suggestions are offered for theoretical and empirical development.

What drives the purchase of high technology? Although the easy answer is that it is purchased like anything else, we suspected that the answer might well depend on the categories of technology being considered. This study examines the procurement process for major weapon systems development. For the reasons discussed below, we felt that the two more usual purchase models – diffusion of innovation and marketing – were insufficient to describe a very complex process. Instead, we anticipated that the broadest model might be one borrowed from the organizational strategy literature – and modified to fit the perceptual and influence processes seeming to take place in the unusual environment of weapons procurement. This study is an investigation of the resulting procurement model and its fit – it is not intended as an experimental test of one potential model against another. Nonetheless, the apparent success of the model suggests that it may have applications elsewhere in explaining how less specialized advanced technological systems are procured.

* The authors appreciate the comments of Darold Barnum and Henry Lowenstein to an earlier draft of this article.

A substantial literature concerns the diffusion of innovation – both of new products and of new ideas (cf. *Rogers* 1983, *Rogers* and *Kim* 1981, *Zaltman* and *Duncan* 1977). According to *Rogers* (1983: 164–186), the process by which an innovation is decided upon requires five discernable stages. The first two of these stages, "knowledge" and "persuasion," provide the input for the initial decision process. Knowledge is concerned primarily with characteristics of the decision-making unit, including socioeconomic, personality, and communication variables. Persuasion, in turn, is determined by the innovation's characteristics, including relative advantage, compatibility, complexity, trialability, and observability.

Although controversy concerning *Rogers'* theory continues in the marketing literature, it is relatively simple to fit the introduction of high technology consumer products such as laser discs or capital goods such as copying machines to the model. More difficult, however, is any effort to explain the development and procurement of major high technology weapon systems primarily restricted to government purchase. The persuasion characteristics of trial and observation may not be available before a development commitment must be made; the knowledge-related variables such as personality and adequacy of communication often cannot be formally acknowledged as a part of the decision process.

In contrast to diffusion theory, strategic organizational pressure may play an important but generally unexamined role in the process of weapon systems acquisition (*Fahey* and *Christensen* 1986). The primary purpose of this research was to examine the relationship of strategic manipulation to the procurement process. Of secondary interest were differences between competing groups in their perception of strategic manipulations, and the relationship between organizational commitment and success with the procurement contract. Final parts of this article discuss the practical and theoretical implications of our findings.

Research Issues

Over $ 94 billion of the fiscal 1984 U.S. budget was allocated to the purchase of major weapon systems; $ 663 billion in spending was planned for the subsequent five years, with the bulk of that money earmarked for fewer than fifty high cost systems. Winning or losing a single contract in this highly competitive market can mean a difference of several billion dollars in annual sales to a prime contractor and can force significant shifts in organizational strategy. The contractors themselves are largely dependent on defense business: excluding IBM, the 18 largest defense contractors (winning 1982 awards in excess of $ 1 billion) relied on the Department of Defense for 47.2% of their business (*Baumgartner* 1983). One might anticipate that contractors facing pressures of this magnitude would choose any ethical strategy by which they might manipulate decisions to their advantage – and perhaps accept strategies of questionnable standards as well (*Horrock* and *O'Shea* 1985).

Defense contracting organizations negotiate with a monopsonistic customer complex (their government) and the behavior of that customer must be correctly understood. On its face, the problem of the contractor is straight forward: it must convincingly meet the needs of the government (on time delivery of a high quality product at a low price). For the reasons outlined below, we predicted that this needs analysis is overly superficial. Instead, we adapted and examined the strategic manipulation dimensions suggested by *MacMillan* (1978) in order to explain the procurement of major weapon technology.

Strategic Manipulation

In his description of the means by which an individual might manipulate a situation, *MacMillan* (1978: 13–15) posited two independent strategic dimensions: channel (situational or intentional) and mode (positive or negative). In order to describe defense contractors attempting to obtain strategic advantage, we adapted the *MacMillan* two-dimensional model to potential supplier-customer manipulation (98–101). We anticipated that a manipulation might be *situational* (with the supplier altering the apparent situation) or *static* (with the supplier encouraging acceptance or rejection of the existing situation). The second strategic manipulation dimension concerns the result: there may be a *reward* anticipated by the customer if the supplier obtains the contract, or a *punishment* anticipated if the contract is instead awarded to a competitor.

In combination, these two dimensions yield four quadrants for potential strategic manipulation (Table 1). An apparent situational cause, in combination with the suggestion that the customer will be rewarded if the firm is selected, can be termed an inducement (such as low cost). The same cause, combined with the threat of punishment if the firm is not selected, can be termed a coercion (suggesting that the project will fail). A static cause combined with the possibility of reward leads to persuasion (such as Congressional funding if the project is accepted). Finally, the combination of a static cause with the possibility of punishment could be termed obligation (retired friends might be hurt). The nature of our expectations for these quadrants is discussed more explicitly in the paragraphs below.

Inducement. Good performance, low cost, and delivery (the first three traditional "Ps" of marketing) can each be described as "inducements" in *MacMillan's* model; they are controlled by the supplier and can be used to improve the situation. Each might be assessed by a potential customer in terms of the supplier's prior track record. The bid price can also be used to assess a potential

Table 1: Potential Methods for Interorganizational Strategic Manipulation

	Apparent Cause	
Result	Situational	Static
Customer Will Be Helped If Manipulator Is Selected	Inducement 1. Technical track record 2. Schedule track record Cost considerations 3. Track record 4. Final cost 5. Bid price 6. Buy-in	Persuasion 9. Congressional influence 10. Single issue
Customer Will Be Hurt If Manipulator Is Not Selected	Coercion 7. Ghost stories 8. Deterioration of industrial base	Obligation 11. Government retirees 12. Early involvement

Note: Categories are adapted from *MacMillan*, 1978. Quadrants (inducement, coercion, persuasion, and obligation) represent the intersection of the overall Result (reward-punishment) with the Cause (situational-static) dimension. Items within quadrants represent survey items designed for this study.

contractor. Nonetheless, analyses of contractor-government relationships suggest that neither side entirely understands what is or should be objectively required in a project (*Babin* 1982: 10–12). Fixed price bidding on large defense development contracts seldom occurs, since the business failure of a contractor would prevent delivery and erode the capital base of important contractors (*Edwards* 1984). Consequently, cost might be additionally evaluated in terms of anticipated final cost and the probability of a "buy-in" in which the contractor offers a low initial price in the expectation that it can somehow recover its costs in later production work or price increases arising from engineering changes (*Gordon* and *Gallagher* 1984).

Because they largely work with a single customer, defense contractors effectively face a controlled market. Overlaying this single customer control are the funding authority and influential pressure of Congress. Although Congressional support affects both the programs and suppliers selected, it also defines the formal conditions of competition. In somewhat similar conditions of regulation, firms producing for the private sector have sought unregulated and sometimes unethical alternative means to compete successfully (*Kotler* and *Achrol* 1984: *Lawyer* 1963, *Miles* and *Cameron* 1982). Similarly, we anticipated that delivery-product-price strategies would account for only a portion of the strategic alternatives used in obtaining defense industry contracts (*Meyers, McClenon,* and *Taylor* 1982: ii). Potential strategies based on the other three quadrants are described below.

Coercion. Although inducement strategies might be seen as more acceptable and expected, the relationship between their use and success is not clearcut

(*Fahey* and *Christensen* 1986), and work on interfirm manipulation suggests that inducements are less likely to be used as a firm's competitive situation deteriorates (*Wilkinson* and *Kipnis* 1978). A firm might prefer "coercion" in which it suggests that the customer will be somehow punished if the firm is not selected as the contractor ... but will avoid the punishment if the firm is selected. Techniques for exercising coercion could include spreading "ghost stories" concerning the incompetence or prior failure of competing contractors (*Beveridge* and *Velton* 1982: 96). If successful, the rumors could make the use of alternative sources seem too risky, forcing the government to contract with the firm initiating the rumors, despite possibly paying some premium for that privilege. Appeals to the need to maintain the national industrial base and future competition by awarding a contract to an otherwise less deserving firm may also be placed in the coercion category. To the extent that the contractor's technological or manufacturing base is seen as being necessary and irreplaceable, it is a strategic resource which may be worth special support (*Larwood* 1984: 330, *Pfeffer* 1981: 101–103).

Persuasion. Without changing the actual situation or its perception, a contractor might affect reaction to it through "persuasion" in which the customer finds potential advantage in selecting the contractor. Interorganizational forms of persuasion between a weapons supplier and the government might include congressional influence. A firm could seek indirectly to pressure procurement officers through members of the Congress whose districts would be affected, or directly to persuade Congress of the importance of continuing program funding (*Oliverson* 1982: 188–189). Alternatively, the firm could seek to focus attention on a single issue in which it holds a competitive advantage (*Beveridge* and *Velton* 1982: 5), thereby encouraging the customer to ignore the firm's overall similarity to other suppliers and any issues at which it is at a disadvantage.

Obligation. Finally, a contractor could show, without changing the situation, how the customer will be hurt by working with an alternative supplier – thereby imposing an "obligation" on the customer to work with it alone. Major weapon systems development projects may take several years of initial exploration; thus early involvement may serve the purpose of locking in or obligating the buyer to a later contract. Of course, early involvement may simultaneously help shape the nature of the resulting request for proposal or RFP (*McDonald* 1978: B-1-20-21, *Oliverson* 1982: 186, 256). Similarly, hiring government retirees may psychologically obligate their former colleagues to help the hiring firm (*Gordon* and *Gallagher* 1984) – although recent regulations limit this strategy.

We recognize that none of the strategies discussed above can be considered pure examples of just one of *MacMillan's* (1978) manipulation quadrants. For example, while the employment of government retirees may help to obligate weapon system purchasers, high status retirees may also serve to persuade the opinions of legislators and other decision makers. This research was designed to apply the *MacMillan* model as an explorative device in examining defense industry strategies. If successful, the research would show the efficacy of management strategies suggested by the model, beyond the already widely known marketing strategies described by the inducement quadrant.

Perceptual Differences and Commitment

We had two additional purposes. First, we intended to compare the perceptions of government officials concerning the use of different strategies with perceptions of industry contract winners and losers. Comparisons are potentially important in understanding the overall responses. It was anticipated that because of their differing perspectives members of the three groups would view use of the strategies somewhat differently. In particular, losers might be expected to degrade the inducements offered by winners and to see alternative manipulations as having been more important. The losers would thereby depict themselves as unjustly hurt by a competitor of dubious integrity – while winners would uphold their own integrity. To the extent that both agree on strategies, however, the result provides evidence that the strategy really is used, thereby strengthening the overall findings.

Second, we examined the extent to which the investment of effort and attention by contractors was thought related to their actually obtaining the contract. In a review of literature concerning strategy under uncertainty, *Wernerfelt* and *Karnani* (1984) pointed out that most analysts recommend lack of commitment and flexibility be adopted as a strategy until the future is clarified. Nonetheless, *Wernerfelt* and *Karnani* maintain that uncertainty may require the immediate focus of efforts to the extent that special economies may accrue to the firm investing first. This assertion concerning the importance of long-term thinking is consistent with many analyses of the Japanese competitive advantage over western firms (*Feigenbaum* and *McCordruck* 1983). Since the technology used in the defense industry is complex and some RFPs are written with the assistance of the contractors most likely to be involved later (cf. *Oliverson* 1982: 186–256), it seemed likely that substantial commitment prior to award of the contract would be associated with a higher probability of obtaining it.

Overview and Research Questions

Recapping the primary issues above, the purposes of this study were to investigate simultaneously the application of *MacMillan's* model to the business of defense contracting, and the manner in which influence is perceived as used in that industry. The model is a potentially important alternative to diffusion of

innovation in particularly restrictive market conditions. In this process, we were led by only one primary hypothesis:
1: We anticipated that the *MacMillan* model, as adapted to this situation, could be used to predict the perception of four forms of strategic influence:
a) inducement
b) coercion
c) persuasion
d) obligation.

Additionally, we had two research-based questions concerning defense contracting which could help shed light on the process:

2: Do those winning contracts, those losing them, and those awarding them perceive the influence process in the same way?

3: How early must commitments be made in order to improve the likelihood of winning defense contracts?

In order to investigate these issues, we surveyed defense industry program managers on both the winning and the losing sides of major development contracts. Respondents assessed influence and investment strategies based in part on the *MacMillan* model. Their responses were compared with one another and with those given by government program managers in order to determine the efficacy of the model and to answer the additional research questions. The implications of our findings for both managers and researchers are discussed in the final sections of this chapter.

Method

Two hundred and fifty managers associated with one or more major weapon acquisition decisions and familiar with U. S. Department of Defense (DoD) procurement policies were invited to respond to a mail survey concerning the decisions. Fifty of the managers were drawn randomly from DoD lists of armed service program managers, 150 were drawn from American Defense Preparedness Association program managers, and a final 50 were selected to represent high visibility programs. Of these, 142 returned completed surveys (57%). This group represented 45 government and 97 industry responses.

Respondents were asked to base their answers on a single competitive development program of their choosing with which they had first-hand knowledge. The modal program chosen had developmental costs exceeding $ 100 million. Although responses were anonymous, participants indicated the nature of the program and their own relationship with it. Within the industry subset, 56 persons had been on the winning side of the contract they evaluated, while 40 had lost (one did not indicate).

The respondents then answered a series of closed-end questions intended to determine the extent to which the potential strategies noted above seemed to have been instrumental in the award of the contract. In order to maximize cooperation, questions were intentionally worded in the jargon of the defense industry based on the first author's experience in defense contracting. The instrument was pretested with an industry sample in order to assure understanding of the questions.

As a result of the pretesting, some questions were dropped or altered. The final instrument consisted of three types of question, those concerning background, strategy, and investment. The full wording of the instrument is available in *Edwards* (1984), and only portions will be repeated here (in Tables 2 and 4). Because of efforts to deliberately word questions in industry jargon, the anchors and scaling frequently vary between questions. Despite the pretest and wording concessions, eight persons wrote to say they found the research area too delicate to respond and are not included here; a question concerning government-contractor relations was dropped from the analysis when 99 of the final participants failed to complete it.

Results

Strategic Manipulations

The primary hypothesis (1) predicted that the four types of strategic manipulations identified by *MacMillan* would together describe the influence process involved in defense contracting. Questions evaluating each of the potential strategic manipulations suggested by the four quadrants of Table 1 are examined in the sections below. The data and statistics are summarized in Tables 2 and 3.

Inducement. Strategic inducements as defined in Table 1 are normally considered the most publicly acceptable means of strategic manipulation. In order to examine their importance in the present situation, respondents were asked to assess the track record of the *winning contractor* with respect to technical skill, cost, and ability to meet schedule. Winners were rated as having above average technical and schedule records but cost records that did not differ from average [respectively t (134) = 13.38, $p <. 01$; t (133) = 2.83, $p <. 01$; t (133) = .37, $p > .10$ on two-tail tests of the difference between the means and the scale midpoints marked "average" – see Table 2].

Respondents also indicated that they felt the winning contractor's "best and final price" was relatively low (t [127] = 2.17, $p < .05$). Nonetheless, in 41 of the 129 cases considered, the winner was believed to have the highest price, while in 47 cases, the winner was thought to have the lowest price. Evaluating the likelihood of the program attaining final cost objectives, participants pro-

Table 2: Evaluations of Winner's Strategic Manipulation

Strategy	Mean	S. D.	N
Inducement	3.99**	.86	135
1. Winning contractor's track record in terms of technical performance: poor (1) – *average (3)* – superior (5)			
2. Winning contractor's track record in terms of schedule performance: poor (1) – *average (3)* – superior (5)	3.23**	.88	134
3. Winning contractor's track record in terms of cost performance: poor (1) – *average (3)* – superior (5)	2.97	.94	135
4. What was the best and final price position: lowest price (1) – lower half (2) – *2.5* – upper half (3) – highest price (4)	2.26*	1.25	128
5. Projection of program obtaining cost objectives: significant underrun (1) – *target cost +/– 10% (2)* – overrun by more than 30% (4)	2.99**	.85	135
6. Did the Government's procurement policy encourage or result in an overall underpricing of "buy in": clearly evident (1) – some evidence (2) – probably not (3) – no basis for determination (4)	2.25na	.74	108
Coercion			
7. What use did the winning contractor make of "ghost stories" to discredit the competition: negative impact (1) – *not a factor (2)* – highly effective (4)	2.22**	.59	116
8. Did preserving the national technical/industrial base influence the outcome: *not major issue (1)* – major consideration (2) – determinig factor (3)	1.20**	.45	137
Persuasion			
9. Was Congressional influence a factor: major factor (1) – exerted to no effect (2) – negated by opposing pressure (3) – *no pressure (4)*	3.63.**	.79	137
10. Was the selection based on a single key issue: no (1) – *not sure (2)* – yes (3)	2.01**	.84	134
Obligation			
11. What effect did employment of retired government key personnel have on the outcome: negative (1) – *no apparent effect (2)* – key element in winning (4)	2.23**	.62	138
12. When did winning firm become actively involved with this specific program: 3 to 12 months before RFP (1) – > 1 year before RFP (2) – conceived by contractor (3) – follow-on (4)	2.38na	1.06	141

Note: Means marked by ** differ at .01 (* differs at .05) or better by a two-tail *t*-test from the anchor in italics. Note that scale midpoints sometimes differ from the test anchor due to independent item scaling; means marked "na" were not scaled appropriately for testing.

jected an eventual cost overrun of 10 to 30% (t (134) = 13.53, $p < .01$ in a test of the difference between the mean and the anchor "target +/− 10%"). Finally, 19 respondents felt there was clear evidence of a buy-in. Overall, these results show considerable agreement that strategies involving technical expertise, delivery schedule, and cost are believed to be used by contract winners.

Coercion, Persuasion, and Obligation

The two coercion strategies examined were use of ghost stories and arguments that failure to obtain the contract will threaten the national industrial/technical base. Respondents indicated a belief that ghost stories had been helpful to the winner (t [115] = 4.23, $p < .01$), and also felt that concerns for the national industrial base had played a role in selection of the contract winner (t [136] = 5.20, $p < .01$].

In the examination of persuasion strategies, the mean for Congressional influence differed significantly from the neutral option (t [136] = 5.48, $p < .05$), but suggested that most of the pressure was self-cancelling and thought to be of no ultimate benefit. Participants were also asked whether they felt that the final decision was based on a single issue as a competitive discriminator. The mean did not differ from the scale midpoint (not sure) but was significantly above the negative option (t [133] = 14.09, $p < .01$), indicating some doubt as to whether a single issue strategy was worthwhile.

Finally, the two obligation questions were examined. Since virtually all defense contractors employ key retired government personnel to some extent, a question was asked concerning their effectiveness in gaining the contract. Overall responses were significantly more positive than the "no apparent effect" anchor (t [137] = 4.36, $p < .01$). The timing of initial program involvement by the winner was seen as most often beginning at least a year before the RFP (68 responses), or resulting from a prior contract.

Overall, the pattern of the results for the coercion and obligation strategies indicates that these are considered to be important to firms in gaining defense contracts. Some evidence also supported the use of persuasion strategies. Combined with the results in the previous section, evaluating the apparent importance of inducement strategies, these results both support the *MacMillan* model as predicted by our hypothesis, and suggest that reference to inducement strategies alone is not sufficient to explain defense procurement decisions.

Perceptual Differences

One of the intentions of this study was to determine whether perceptions of what is needed to win a contract differ or are similar for respondents in different positions (research question 2). The pattern is potentially important in understanding the overall responses. For example, if winning contractors saw themselves as using only inducement strategies, and losing contractors viewed winners as using only coercion, obligation, and persuasion, the resulting pattern might be assumed to be more the result of impression management or of sour grapes than of factual reporting.

The three respondent groups were compared using two-tailed t tests. As indicated in Table 3, significant differences were found in evaluations of four inducement questions. Losing contractors disagreed with both government respondents and winners, feeling that the winner's technical track record was less successful and that the projection of final cost yielded a higher overrun. Despite these differences, it is important to note that all groups agreed that the winner's technical track record was above average and that an overrun occurred. In the third inducement difference, losers felt that the winner's on time record was relatively poor, while winners felt that it was relatively good. Finally, losers were more likely than government project managers to feel that winners had bought in.

Table 3: Comparisons of Strategic Perception

Strategy	*by Government* Mean	S. D.	N	*by Winners* Mean	S. D.	N	*by Losers* Mean	S. D.	N
Inducement									
1. Technical track record	4.09A	.87	45	4.18A	.85	54	3.56B	.77	36
2. On time record	3.16AB	.83	44	3.52A	.84	54	2.89B	.89	36
5. Final cost projection	2.77B	.87	43	2.84B	.76	55	3.46A	.77	37
6. Did winner buy in	2.38A	.74	34	2.32AB	.72	41	2.03B	.73	33
Coercion									
7. Use of ghost stories	2.03B	.43	39	2.43A	.71	49	2.11B	.42	28
8. Threaten ind. base	1.21AB	.54	43	1.28A	.49	54	1.08B	.27	40
Persuasion									
10. Single issue selection	1.87B	.89	39	1.89B	.83	55	2.32A	.73	40
Obligation									
11. Hire retired personnel	2.23AB	.53	43	2.09B	.62	55	2.42A	.68	40

Notes: Full questions and anchors are in Table 2; only items with statistically significant group differences appear here. Within any row, group means having different coefficients differ .05 or better by a two-tailed t-test. For example, A differs significantly from B, but neither differs from AB.

Considering the coercion strategies, winners felt that ghost stories were more important than either government or losing respondents. Winners also felt that threats to the industrial base were more important than believed by losers. Examining the persuasion strategies, losers felt that selection was more likely to be based on a single issue than did either government or winning respondents. Winners were marginally ($p < .10$ by a two-tail test) more likely than government respondents to report no Congressional influence. Nonetheless, four respondents, including three government project managers, felt that such influence was a major factor in the contract award; in subsequent interviews, several respondents stated that such influence is more important in protecting a program's later funding priority than in determining which firm wins the contract. Among the obligation strategies, losers were more likely than winners to see retired government personnel as having influenced the result.

These results show clearly that there are differences in the manner in which the three groups perceive procurement strategies to have been employed. Nonetheless, the differences do not appear necessarily to assume a self-serving form. For example, winners admitted to the importance of coercion strategies, while losers generally agreed on the importance of the inducement strategies even while they devalued them somewhat.

Commitment

Our final research question (3) concerned the extent to which early commitment is helpful in obtaining contracts. The finding for obligation strategies, that program involvement begins early, is consistent with an industry in which developments must of necessity proceed over a period of years. In order to further examine the extent and effect of early commitment and preparation, we asked contractors to evaluate their own effort and attention to the program in a short series of questions (Table 4). It should be noted that these questions differ from the series reported above in that losing contractors evaluated *their own* firms.

Contractors generally made a direct investment in the competitive effort of between \$ 1 and 5 million (mean 2.33 on a scale from 1 [\$ 0 to 1 million] to 4 [over \$ 5 million]). Government project managers overestimated the amount spent by winners ($p < .05$ by a two-tailed t test of the difference in means), winners and losers did not differ reliably from one another. However, when asked to compare the winner's investment to that of the closest competitor, both government respondents and winners said that winners had invested significantly more than average (for government and winners respectively, t [10] = 2.18, $p < .05$; t [52] = 4.45, $p < .01$ in a two-tail test of the difference between the means and the neutral "about equal" midpoint of the scale). In contrast, losers stated that their own investment was only about average – and significantly less than winners gave for themselves ($p < .05$).

A similar pattern was found in responses to a question asking how top management regarded the competitive opportunity (scaled from 1 [very little interest] to 5 [must win program]). All three groups scored management as giving the contract reliably ($p < .05$) more than the midpoint "balanced interest" (overall mean 4.06); nonetheless, both government and winners rated the interest of the top management of winning contractors above the rated interest of top managers in losing firms ($p < .05$). Finally, all groups felt that the program was a good match with the firm's existing business and experience [mean 3.11 on a scale ranging from "poor match" (1) to "near perfect" (4)]. In sum, both winning and losing contractors showed substantial commitment to the program, but a higher level was often shown by winners – a level corroborated by the government project managers.

Table 4: Evaluations of Contractor's Commitment and Preparation

Investment	*of Winners by Government*			*of Your Firm by Winners*			*by Losers*		
	Mean	S. D.	N	Mean	S. D.	N	Mean	S. D.	N
Total direct investment $0 to $1 million (1) – $5 million (4)	3.38A	1.09	16	2.20B	1.09	56	2.08B	1.14	40
Compare to competitor much less (1) – much more (5)	3.54AB	.82	11	3.55A	.90	53	3.00B	1.09	38
Top management view very little interest (1) – must win (5)	4.29A	.64	21	4.27A	.82	56	3.65B	.95	40
Match to existing business poor match (1) – near perfect (4)	3.15	.72	27	3.07	.74	56	3.13	.76	40

Note: Within any row, subgroup means having different coefficients differ from one another at .05 or better by a two-tail *t*-test. For example, A differs significantly from B, but neither differs from AB. Testing of overall means against scale base anchors was inappropriate for these items.

Discussion

This study has examined the strategies determining high technology weapon systems development. As noted in the introduction, the diffusion of innovation model cannot readily be extended to the acquisition of weapon systems, because some of the elements necessary for the model are missing. The findings concerning strategic manipulation provided clear evidence that technical and delivery records, as well as cost, are important to the acquisition of weapon systems. This part of the findings might be predicted from a straightforward marketing model. Because defense contracting limits the nature of promotion, but clearly defines the market and locks the customer and government into long term relationships with one another, we anticipated the need for a broader model than that commonly used in marketing. *MacMillan's* strategic manipulation model posited that marketing "inducements" would only partly account for a winner's success. Strategies involving persuasion, coercion, and obligation might also contribute to it.

As operationalized here by means of a survey of both government and firm project managers, the strategic manipulation model was generally supported, providing a good fit to hypothesis 1. In addition to the inducement strategies, the project managers surveyed indicated that manipulation by means of coercion (such as passing along ghost stories concerning competitors' incompetence) and obligation (such as hiring government retirees) strategies also took place in important weapon systems development contract decisions in which they were involved. There was some weaker evidence supporting the use of persuasion strategies (using Congressional clout, for example) as well.

Does this show that contractors are unnecessarily manipulative or that only contractors who are wise in the ways of organizational pressure politics can succeed with their technologies? Survey questions which were phrased so as to examine the possibility that a particular strategic manipulation was decisive in the acquisition process did collect some affirmative responses. Similarly, a question asking whether the decision had been based on a single issue (without specifying the issue) found agreement that there was some possibility of this. The substantial variance for all questions, however, indicates that any particular issue, however important, was unlikely to be sufficient. Apparently inducement, persuasion, coercion, and obligation strategies can each help a contractor, but are seldom sufficient to determine success in obtaining the contract by themselves.

A number of significant differences were found between the respondent groups. The pattern of these differences suggests that contractors who lost an important bid saw winning contractors as offering less in the way of real inducements (ability and price) but more illusory cost inducements (cost overrun and buy-in). Coercions were considered more instrumental by winners than by losers. Losers considered persuasion based on a single issue and obligation based on hiring retired key government personnel as more important than did winners. Since the preparation of bids is expensive, it seems unlikely that a losing project manager would bother to bid if he or she knew that the award decision would be based on criteria impossible to meet. Thus we tend to give more credence to the responses of winners and government project managers. Admission of the use of ghost stories by winners may be seen as a particularly strong case in this regard: The devious nature of this coercion suggests that it would be under-reported by those using them. Thus winners reporting them are almost assuredly accurate.

It is an old saw in management that commitment is important to success. In order to assess the role that commitment may have played in contract awards, we asked a number of follow-up questions. Both winners and losers evidenced a substantial degree of commitment, with mean out-of-pocket investments exceeding three million dollars. There were some seemingly important differences, however. Top management of winners took a greater interest in the program, and the winners felt that they had made a comparatively greater investment in preparing their bids than had their competitors. Since the actual investment for the two groups was similar, the difference in comparison to competition suggests that winners more carefully targeted their investments by spending more than competition on preparing bids they expected to win and perhaps less on projects that were considered long shots or less consequential. Alternatively, the investment of losers may have appeared more diffuse to others, indicating that it was less often spent in a manner visible to others outside the organization.

Despite the levels of significance reported in this study and the conclusions above, the results must be seen as exploratory. The applicability of the findings depends in part on the honesty of survey participants in responding to a very sensitive subject. Unlike most research, the present study examined behavior of questionable ethics and substantial import both to the respondent's career and to national defense procurement policy. There is gathering evidence that anonymous self-report data on questions of ethical behavior is sufficiently accurate for analysis (cf. *Clinard* 1983: 33, *Sonnenfeld* 1981). Nonetheless, both response biases and strong opinions are to be expected from those choosing to answer the survey.

Implications for Managers

From a practical point of view, there are many issues still unresolved. Assuming that the results can be corroborated by others, they are most directly applicable to large systems defense contracting. Within that arena, the results indicate some of the areas in which political strategies may be of use, where they will undoubtedly be used by some competitors, and without which the firm discards potential competitive advantage. It should be noted, however, that those bidding as prime contractors for large weapon systems development are themselves large firms with a large portion of their sales made to the government. We do not expect that smaller firms, or those with a higher proportion of nongovernment customers, would feel the same pressure to use non-inducement strategies. If their competitors similarly restricted the range of their strategies, coercion, persuasion, and obligation strategies might not be needed; on the other hand, they would confer a greater advantage on the firm breaking the restriction.

How applicable are the results to non-defense firms building technology for other firms or directly for the consumer market? Some of the strategies examined here are not immediately applicable. For example, hiring government (or any) retirees is unlikely to influence many outside government contracting. Nonetheless, we suspect that there are analogs which are just as potentially important. As this is written, AT&T has hired an Olivetti executive to run its personal computer operations with the expressed hope that he will reinstill confidence if not direction. Ghost stories are the routine fare of television commercials, and recent ones depict competing main frames as breaking down without hope of repair. Threats by Japanese high tech firms to the North American and European industrial bases are routinely verbalized by North American and European competitors. In short, we believe that the results are, overall, very applicable.

Implications for Theory Development and Further Research

Although the diffusion of innovation model accurately describes the spread of innovation, it provides a poor fit for those seeking competitive advantage over others when the decision is major or pieces cannot be pretested and satisfactorily examined. The marketing model is restricted and can be subsumed under the strategic manipulation model used here. Nonetheless, there are several important requirements for further developing the present model. Ideally, there will be replications under varied conditions. While we have suggested broad applicability above, that is still to be shown. It is of interest to show that the *MacMillan* model is useful in non-defense technology decisions – such as in the contracting of commercial aircraft, and to determine whether it has any utility not already provided by other models in retail technology purchase decisions.

We would like to see a different starting point used as well: other researchers might examine an extended series of strategies, applying factor analysis to determine empirically how they relate with one another. The dimensions of *MacMillan's* model are logically rather than empirically determined. Although we have applied *MacMillan's* model here, we doubt that there is any magic to the four quadrants of that model beyond its usefulness in stimulating consideration of strategic manipulation.

Finally, it would be useful, perhaps through a cluster analysis of industrial and innovation categories, to determine whether some types of strategies are more useful in particular circumstances than others. Ghost stories seem more likely to rely on situations with imperfect (rumor) information transmission, and concerns about erosion of the industrial base seem most likely to occur under isolated or sole-source conditions. Perhaps the entire coercion quadrant relies on imperfect conditions, while other quadrants rely on other types of conditions.

We see this study as a new start, not a conclusion, of research activity designed to determine what work in marketing technology succeeds, and what fails. Realistically the success of even the most exciting technology depends on the strategies selected by those managing its introduction. This research is as crucial as the technology itself.

References

Babin, J. C. (1982): *Federal source selection procedure in competitive negotiated acquisitions*, Washington, DC: George Washington University, National Law Center. (Unpublished masters thesis.)

Baumgartner, J. S. (1983): The top earners among defense industry, *Program Manager*, September–October, 47.

Beveridge, J. M. and *Velton, E. J.* (1982): *Positioning to win*, Radnor, PA: Chilton Book Co..

Clinard, M. B. (1983): *Corporate ethics and crime*, Beverly Hills, CA: Sage.

Edwards, F. L. (1984): *Factors affecting the competitive process in the acquisition of major weapon systems*, Claremont, CA: Claremont Graduate School. (Unpublished doctoral dissertation.)

Fahey, L. and *Christensen, H. K.* (1986): Evaluating the research on strategy content, *Journal of Management*, 12: 167–183.

Feigenbaum, E. A. and *McCorduck, P.* (1983): *The Fifth Generation*, Reading, MA: Addison-Wesley.

Gordon, G. and *Gallagher, A.* (1984): Has Pentagon lost war with defense suppliers? *Chicago Tribune*, December 20: 25, 44.

Horrock, N. and *O'Shea, J.* (1985): Anatomy of a scandal: Firms build immunity to regulation, *Chicago Tribune*, December 22: III, 11, 14.

Kotler, P. and *Achrol, R. S.* (1984): Marketing strategy and the science of warfare, in: R. B. Lamb (Ed.), *Competitive strategic management*, Englewood Cliffs, NJ: Prentice-Hall.

Larwood, L. (1984): *Organizational behavior and management*, Boston: Kent Publishing.

Lawyer, J. Q. (1963): How to conspire to fix prices, *Harvard Business Review*, March–April, 41 (2): 95–103.

MacMillan, I. C. (1978): *Strategy formulation: Political concepts*, St. Paul, MN: West Publishing.

McDonald, P. R. (1978): *Prime contracts and subcontracts service*, Covina, CA: Procurement Associates.

Meyers, M. G., McClenon, P. R., and *Taylor, H. M.* (1982): *Price competition in the DoD*, Washington, DC: Department of Defense, Contract #MDA 903-81-6-0166.

Miles, R. H. and *Cameron, K. S.* (1982): *Coffin nails and corporate strategies*, Englewood Cliffs, NJ: Prentice-Hall.

Oliverson, M. (1982): *Defense marketing management*, Scottsdale, AZ: Oliverson.

Pfeffer, J. (1981): *Power in organizations*, Marshfield, MA: Pitman Publishing.

Rogers, E. M. (1983): *Diffusion of innovations*, third ed., New York: Free Press.

Rogers, E. M. and *Kim, J.-I.* (1981): The diffusion of innovations in public organizations, in: R. L. Merritt (Ed.), *Handbook on the comparative study of innovation*, New York: John Wiley & Sons.

Sonnenfeld, J. (1981): Executive apologies for price-fixing: Role biased perceptions of causality, *Academy of Management Journal*, 24: 192–198.

Wernerfelt, B. and *Karnani, A.* (1984): Competitive strategy under uncertainty, *Academy of Management Proceedings*, 42–46.

Wilkinson, I. and *Kipnis, D.* (1978): Interfirm use of power, *Journal of Applied Psychology*, 63: 315–320.

Zaltman, G. and *Duncan, R.* (1977): *Strategies for planned change*, New York: John Wiley & Sons.

Chapter 4
The Context of Innovation: Investment in R&D and Firm Performance

Michael A. Hitt, R. Duane Ireland, and Igor Y. Goryunov

Abstract

Results show that the degree of a firm's diversification moderated the relationship between R&D intensity (R&D expenditures as a percent of sales) and performance (market returns). Growth strategy and industry type did not affect this relationship. A critical finding showed that negative relationships between R&D intensity and performance existed for firms that diversified beyond their core business but in related products. Most of the significant relationships concerned the related diversified firms. No positive relationships between R&D intensity and performance were discovered in this study. This surprising outcome may be extremely important. R&D investments should contribute positively to firms' performance. If this does not occur, resources are not being managed efficiently and/or effectively. These results have significant implications for the nature of innovation in and global competitive stature of firms in the U.S. and other Western Nations.

Introduction

Hambrick and *MacMillan* (1985) noted that some "Skeptical executives view R&D as the sink hole of the 1980s." They suggested that research results regarding the payoffs of R&D are inconclusive. Further, they argued that negative executive attitudes will persist until more is known regarding factors leading to greater effectiveness of R&D expenditures. *Dewhirst* (1981) suggested that our failure to examine the importance of the research and development function in the study of strategic management may account for a portion of this lack of understanding.

These conditions notwithstanding, the importance and complexity of research and development activities are generally accepted. *Cooper* (1984c) observed that "the importance of new products and technological innovation as a route to corporate growth has long been recognized" (p. 247). The "process" of R&D efforts is also valuable. For example, *Hartwick* (1984) proposed that a firm's ability to win a future "R&D race" is enhanced simply through participation in previous races. Positive benefits accrue to the firm in spite of "losses" in other R&D efforts. *Fine* (1983) suggested that firms producing high quality products may expect their production costs to decline more rapidly than firms producing low quality ones. Thus, R&D efforts to help a firm produce quality products may improve the firm's financial performance.

These findings are interesting. However, the precise context and linkage of innovation to performance improvements have not been specified. The linkage is complicated and revolves around a wide range of issues. For example, *Iyer* and *Ramaprasad* (1984) suggest that a firm's strategic posture toward innovation significantly affects its ability to respond quickly and appropriately to rapidly changing environmental conditions. Strategic posture is defined in terms of "[...] the cost (the firm) is willing to incur, the benefits it seeks, and the risks it is willing to take in each of three phases of innovation" (p. 90). The context of innovation is also affected by a firm's strategy, the primary industry in which it competes, and its level of diversification.

These important issues are examined in this research to enhance our understanding of the context of innovation. They are evaluated through a focus on the relationship between investment in R&D and performance of large U.S. industrial companies. A brief discussion of the importance of the research and development function, and its recent level of support in American companies, precedes the description of the research and discussion of the results.

The Current Setting of R&D

Innovation is critical for firms to maintain their global competitive stature. Research and development activity must contribute positively to the firm's overall performance, or problems will ensue. In the latter 1970's and early 1980's, a number of U.S. firms encountered performance difficulties and lost portions of their market to international competitors. Some (e.g., *Dickson* 1986) have suggested that the fundamental course of international corporate competition has been altered.

In the late 1960's and 1970's, a "crisis of innovation" developed in which less technological progress occurred (*Hull, Hage* and *Azumi* 1985). During the years 1966–1978, R&D expenditures in the U.S. declined consistently as a percentage of GNP. In addition, the number of patents issued in the United States declined by 46.2 percent during the years 1971–1979 (*Science Indicators* 1983). However, since 1979, double-digit increases in R&D outlays have been recorded in U.S. industry (*Business Week* 1986a: 134–138).

While several explanations for the crisis of innovation are possible, *Hayes* and *Abernathy* (1980) argued that many U.S. executives were not willing to accept the risks of innovation because of the lack of predictability of consumer preferences. Thus, firms found it easier to introduce a new version of an existing product (e.g., a slightly modified version of an existing toothpaste) than to truly innovate. As such, many R&D expenditures focused on research for short run success, perhaps at the expense of long run effectiveness. A negative by–product is the tendency for U.S. firms to retreat to relatively safe domestic markets. General Electric's acquisition of the RCA Corporation has been criticized on this ground. Specifically, some analysts suggest that a key reason for this acquisition may be to entrench GE's position in relatively safe domestic markets. *Nussbaum* proposed that

> "Instead of manufacturing for the global marketplace, they (GE) are increasingly seeking shelter in niches of the domestic economy relatively free from foreign competion – defense and services" (*Business Week* 1985: 50).

Ohmae (1985) does not evaluate these kinds of actions favorably. He proposed that competitive success in the future will require market shares within the U.S., Japanese, and European markets. Further, he argued for joint R&D ventures between firms from different countries. Some key decision makers are apparently responding to this advice. Now often labeled "strategic alliances" rather than joint ventures, it was reported recently that firms such as General Electric, Xerox, General Motors, IBM, and AT&T have formed partnerships with firms in Japan, Taiwan, Italy, Spain, South Korea, and the Netherlands (*Business Week* 1986b: 100–105).

Additionally, it became common practice, during the 1960's and 1970's, to "borrow" research outcomes of other organizations instead of investing in R&D [referred to as the Not-Invented-Here (NIH) syndrome]. This resulted in a shift of R&D risks outside the firm and in the creation of a significant dependency relationship. *Statman* and *Tyebjee* (1984) propose that a single unit of analysis (i.e., the project manager) has typically been used to consider the issue of risk. They concluded that the risks of investment in technological innovation differ among the perspectives of individual project managers, upper level managers, and investors. Thus, the relationship between risk and return may be more complex than is usually believed. In turn, decisions during this time period not to invest in a firm's innovation capabilities, due to perceived levels of risk, may have been, in certain cases, inaccurate.

The management of R&D and the amount of resulting innovation has become a significant issue in industry in recent years (*Linn* 1983). However, little research has focused on this matter. Nonetheless, it is important to investigate the relationship between the relative investment in R&D and firm performance.

Corporate Innovation Context

McGinnis and *Ackelsberg* (1983) argue that American corporations must rekindle their innovative and creative skills in order to achieve global competitiveness. *Avard, Catto* and *Davidson* (1982) argue that this relationship is especially critical for large industrial firms. *Levine* and *Yalowitz* (1983) suggest that companies' operations will be affected in meaningful ways by technological innovations during the 1980's and beyond. These effects, they argue, should be anticipated to develop appropriate responses. *McGinnis* and *Ackelsberg* propose that a firm's innovative capabilities should be viewed as a critical component of the strategic planning process. An analysis of the challenges faced by research-intensive firms operating in turbulent environmental conditions led *Kimmerly* (1983) to conclude that incorporating research and development issues into strategic plans is necessary for successful performance. Thus, the linking of R&D and strategy may be critical for a firm's long-range performance.

Strategy

Cooper (1984c) found that firms that carefully conceived a new product strategy performed better than firms that failed to do so. *Cooper* noted that the best performing firms

> "[...] featured a strong, aggressive technological orientation; they were strongly R&D oriented, proactive in acquiring new technologies, and boasted a venturesome, offensive program that was (on the) leading edge of corporate strategy."

Donaldson and *Lorsch* (1983) argue that most firms prefer an internal growth strategy to reduce their dependencies on debt and equity capital markets. Further, they suggest that building superior R&D skills is necessary for firms seeking internal as opposed to external acquisitive growth. However, *Hitt, Ireland* and *Stadter* (1982b) found that executive perceptions of the strategic importance of R&D were negatively related to financial performance measures for firms implementing either an internal growth or an external growth strategy. In an extension of their earlier work, *Hitt* and *Ireland* (1985) confirmed the negative relationship between the relative emphasis on R&D and firm performance. They found that R&D efforts to improve current products and manufacturing efficiency contributed strongly to the negative relationship.

Although limited, the research does suggest a relationship between innovative activity, strategy and firm performance. However, the results are inconclusive with respect to the direction of the relationship. This particular study focuses on corporate level strategies, as measured by a firm's growth strategy.

Industry

MacMillan, McCaffery and *Van Wijk* (1985) found each industry to be idiosyncratic in responses to the introduction of new competitive products, suggesting industry differences in approaches to R&D. Further, *Baysinger* (1985) suggested that aggressive innovation was relatively more attractive in some industries than others. *Hitt* and *Ireland* (1985) found a negative relationship between the emphasis placed on R&D and performance for firms in consumer nondurable goods industries. Finally, *Miller* and *Friesen* (1986) were unable to find support for *Porter's* generic strategies in the cases of firms competing in consumer durable goods industries. This is in contrast to *Hambrick, MacMillan* and *Barbosa's* (1983) results. They found stable strategies (both differentiation and overall cost leadership) for firms competing in capital goods industries. These relationships have been found in manufacturing and nonmanufacturing firms operating in the for profit sector. Thus, a relationship between innovative activity, dominant type of industry and firm performance may exist. However, the research is too limited to specify the exact nature of this potential relationship.

Diversification

Nelson (1959) and *Hambrick* and *MacMillan* (1985) argue that diversified firms can gain greater benefits from R&D because they can better utilize unintended outcomes in diverse units. However, *Kelly* (1970) found that firms with diverse but technologically related products received the greatest outcomes from R&D investments. *Hambrick* and *MacMillan* (1985) found that the relative breadth of a business product line was positively related to innovative efficiency. They examined R&D expenditures at the business unit level. At the corporate level, the same benefits may accrue for R&D outcomes across businesses, but greater coordination costs are incurred in attempts to derive such benefits. Thus, the effects on overall firm performance may not be as significant.

Research Issues

In summary, past research has established that innovation is critical for firms to maintain a globally competitive stature. Research and development activity must contribute positively to the firm's overall performance. However, additional research is required. For example, the precise context and linkage of innovation to performance improvements have not been specified. The literature, although limited, suggests a relationship between innovation activities and performance. Further, the literature suggests that this relationship may vary across different strategies, types of industry and levels of diversification, leading to the following research questions.

Research Question 1: What is the relationship between total relative corporate R&D investment and corporate performance in firms using different strategies?

Research Question 2: What is the relationship between total relative corporate R&D investment and corporate performance for firms in different types of industry?

Research Question 3: What is the relationship between total relative corporate R&D investment and corporate performance in firms with different levels of diversification?

Method

Sample

The data were gathered for 124 firms listed among *Fortune's* list of 1000 industrial companies. Data, regarding these firms' grand strategy, major industry type, and degree of diversification, were obtained from questionnaires completed by each firm's chief executive officer and another senior level executive knowledgeable of overall firm operations. Corporate performance and R&D intensity data were obtained from secondary sources described in the sections that follow.

Survey Data

Grand strategy was measured on a nominal scale with four strategy classifications: 1) stability (similar operating levels to pursue similar objectives with only incremental performance improvements); 2) internal growth (growth through internal development); 3) external acquisitive growth (growth through acquisition, merger and/or joint ventures); and 4) retrenchment (reduction of level or scope of product/market objectives).

A second nominal scale was used to measure industry type. Four normative classifications were employed: 1) consumer durable (goods purchased for personal gratification that last for long periods of time – automobile); 2) consumer nondurable (goods purchased for personal gratification but last for only a limited time period – processed foods; 3) capital (goods used to manufacture or provide consumer or producer goods to others – machinery, buildings); and 4) producer (goods used as raw material to produce consumer or capital goods –

steel, cement). Executives noted their firm's *principal industry* (defined as the one in which the largest percentage of sales is earned). This classification scheme, derived from *Economics Encyclopedia* (1973, 1974), *Khandwalla* (1977), and *Schoeffler, Buzzell,* and *Heany* (1974), is similar to the one used by *Snow* and *Hrebiniak* (1980) to classify firms and is utilized extensively by practitioners. Nonetheless, this measure is a broad classification scheme and as such, may not fully account for differences among specific industries (e.g., automobile manufacturers and manufacturers of PC's both produce consumer durable goods, yet they operate in substantially different industry settings).

An ordinal scale was used to measure diversification strategy. Four strategy classifications, each representing an increasing degree of diversification, were included. These classifications, used originally by *Wrigley* (1970) and later by *Grinyer, Yasai-Ardekani,* and *Al-Bazzaz* (1980), were: 1) single business (95 percent or more of annual sales from one end product business); 2) dominant business (70 to 94 percent of annual sales from one end product business); 3) related business (less than 70 percent of annual sales from one end product business and diversification primarily in concentrically related products); and 4) unrelated business (less than 70 percent of annual sales from one end product business and diversification unrelated to the primary end product business).

Hitt, Ireland and *Palia* (1982a) and *Hitt* and *Ireland* (1985) reported acceptable reliability and validity levels for all three measures.

Corporate Performance

Market returns, as reflected by the Treynor Index (*Treynor* 1965, 1968), was used to measure financial performance. It has been suggested (*Reilly* 1979, *Weston* and *Brigham* 1978) that the securities market adjusts or compensates for deficiencies associated with some accounting data. Thus, the market evaluates a firm's present and prospective earnings flow, the timing of risks associated with this flow, and the firm's dividend policies. In total, the market price of a firm's stock reflects the organization's long-run performance potential on behalf of stockholders, which *Rappaport* (1978) argues is necessary to reflect a firm's performance accurately.

Market returns was calculated as follows:

$$MR = \frac{GMR_i - RFR_k}{\beta}$$

where MR = market returns; β = beta measure of systematic risk; GMR_i = geometric mean annual stock return for firm i; and RFR_k = geometric mean annual riskless rate.

Data on daily market returns for firms' securities were obtained from CRSP tapes. A maximum of five years (1975–1979) data was used to calculate the geometric mean annual return. The geometric annual risk-free rate was computed from the average selling prices at the weekly three-month treasury bill auction. These data were obtained from the *Federal Reserve Bulletin*. The beta measures of systematic risk were taken from the *Value-Line Investment Survey*. The mean betas presented in *Value-Line* are calculated for a period of five years, matching the five years of market returns data.

Market returns has potential limitations. As with any financial performance measure, it can be affected by some extraneous factors (e.g., economic and industry trends). However, a large percentage of the variance in market returns can be attributed to a firm's actions over time. *Branch* and *Gale* (1983) suggested that there are four determinants of stock price: 1) firm profitability; 2) firm growth; 3) risk; and 4) stock market outlook. Two of these (profitability and growth) are affected principally by a firm's actions. In addition, *King* (1966) found only approximately 10 percent of variance in stock prices to be explained by industry. Finally, adjustments for risk accounts for a portion of industry effects on performance. In the current sample, no differences in mean market returns by type of industry were found.

Bowman (1980) questioned the positive relationship between risk and return. However, he focused on accounting-determined return measures (e.g., returns on equity) rather than market returns. *Bettis* (1981) noted that although not conclusive, the overwhelming weight of research evidence suggests that returns should be adjusted for risk. Therefore, this measure has advantages over most other financial performance measures. In summary then, market returns is based on the capital asset pricing model, represents the best theoretically derived firm performance index now available (*Langetieg* 1978) and has become accepted in the finance literature in the study of corporate strategies (*Harrington* 1983, *Lubatkin* 1983).

R&D Intensity

R&D intensity is the amount of research and development expenditures as a percentage of sales. Using this measure, rather than the absolute amount of research and development expenditures, dilutes the effects of size and controls for heteroscedasticity (*Hambrick, MacMillan* and *Barbosa* 1983). Further, a measure of R&D intensity reflects a firm's commitment to research and development activity and permits relative comparisons among firms. Data on total research and development expenditures and sales were obtained from the annual *Business Week Corporate R&D Scoreboard* (1975–1979). The resulting figure reflects R&D intensity for the total corporation and as such, may be termed corporate R&D intensity.

Results

Moderated regression analyses were used to test the moderating effects of strategy, type of industry, and degree of diversification on the relationship between R&D intensity and performance. The results are displayed in Tables 1, 2, and 3. As shown, no strategies nor types of industry were found to be moderators. However, two levels of diversification did moderate the relationship between R&D intensity and performance. Table 2 shows that both the dominant and related business diversification levels were statistically significant moderators. These results suggest that type of strategy and type of industry alone may not be major contingency variables with regard to the effectiveness of R&D investment. These results provide answers to the research questions denoted earlier.

Table 1: Moderated Regression Analysis with Strategy as the Moderator

	Restricted Model	Full Model		
Strategy	R^2	R^2	ΔR^2	F
Stability	.0235	.0303	.0068	.84
Internal Growth	.0091	.0153	.0062	.76
External Growth	.0551	.0564	.0013	.17
Retrenchment	NA	NA	NA	NA

Table 2: Moderated Regression Analysis with Level of Diversification as the Moderator

	Restricted Model	Full Model		
Diversification	R^2	R^2	ΔR^2	F
Single Business	.0107	.0188	.0081	.99
Dominant Business	.0349	.0773	.0424	5.36**
Related Business	.0150	.0610	.0460	5.88**
Unrelated Business	.0075	.0075	.000	0.00

**$p < .01$

Table 3: Moderated Regression Analysis with Type of Industry as the Moderator

	Restricted Model	Full Model		
Industry	R^2	R^2	ΔR^2	F
Consumer Nondurable	.0314	.0439	.0125	1.59
Consumer Durable	.0358	.0505	.0147	1.86
Capital	.0616	.0773	.0157	2.04
Producer	.0387	.0472	.0040	.50

Based on the results of the moderated regression analyses, Pearson product moment correlations between R&D intensity and performance were developed for firms using the different levels of diversification. The results are included in Table 4. As shown, the relationship between R&D intensity and performance was statistically significant only for related business firms. The relationship in this case was negative. The relationship for dominant business firms was positive and approached significance.

Table 4: Correlation Between R&D Intensity and Performance

Diversification	Correlation Coefficient
Single Business (n=4)	.471
Dominant Business (n=26)	.252
Related Business (n=63)	–.301**
Unrelated Business (n=31)	–.060

$p >$ .05**

The previous analyses suggest the strong influence of related diversification on the investment in R&D. However, it may be erroneous to conclude that it is the only contingency factor of those studied influencing R&D investment. *Hitt, Ireland* and *Stadter* (1982b) and *Hitt* and *Ireland* (1985) found both strategy and industry to interact in their effects on the relationship between R&D as a distinctive competence and performance. Therefore, these contingency factors may interact with others to affect the effectiveness of R&D investments.

To examine potential interaction effects among strategy, industry, and diversification, further analyses were conducted. Correlations between R&D intensity and firm performance were calculated for each strategy × industry, industry × diversification, and strategy × diversification cell. The results of these analyses are shown in Table 5.

Table 5: Correlation between R&D Intensity and Performance: Strategy x Industry x Diversification Cells

Firm Characteristics	Consumer Nondur.	Consumer Durable	Capital	Producer	Single Business	Dominant Business	Related Business	Unrelated Business
Stability	−.187 n=5	NA	NA	−.631* n=8	NA	−.484 n=4	−.711* n=7	NA
Internal Growth	.082 n=25	−.379 n=7	−.225 n=28	−.322 n=27	NA	.338 n=20	−.250* n=46	−.115 n=20
External Growth	NA	NA	.334 n=7	.237 n=6	NA	NA	−.044 n=9	.315 n=6
Retrenchment	NA	NA	NA	NA	NA	NA	NA	NA
Consumer Nondurable					NA	.386 n=9	−.192 n=14	−.104 n=7
Consumer Durable					NA	−.677 n=4	NA	−.511 n=4
Capital					NA	.600 n=4	−.564*** n=23	−.167 n=9
Producer					NA	.526 n=9	−.480** n=24	−.078 n=11

***$p > .01$
**$p > .05$
*$p > .10$

The statistically significant relationships between R&D intensity and performance were found in the stability × producer, stability × related business, internal growth × related business, capital × related business and producer × related business cells. Again, each significant relationship was negative. Thus, in these instances, firms that have allocated a significant percentage of sales volumes to research and development activities may experience a decrease in their performance. However, the positive relationship in the internal growth × dominant business and producer × dominant business cells *approached* significance. Some of these results support previous findings (for example, in the stability × producer cell). Since strategy and industry effects were examined in conjunction with degree of diversification in this study, these results may facilitate further interpretation of findings reported previously by *Hitt* and *Ireland* (1985). Most effects reported herein existed in interactions between strategy and diversification and industry and diversification.

Discussion

The Research Issues

The results indicate that in firms that have diversified into related businesses, R&D intensity is negatively related to performance. No statistically significant positive relationships were found. Although counterintuitive, the results support *Tornatzky* and *Klein's* (1982) findings (in three separate studies) of a negative relationship between profitability and adoption of innovation. Strategy and industry effects were present in interaction with other contingency variables, but the dominant influence was related diversification. Negative relationships between R&D intensity and performance were discovered in related business firms following stability or internal growth strategies and those in capital or producer goods industries. Although *Hitt* and *Ireland* (1985) found a negative relationship between R&D and performance for internal growth strategy firms, in this study, such a relationship was found for only internal growth firms that were related diversifiers. Further examination of Table 5 provides information helpful in explaining this result. For internal growth and for producer firms that were dominant diversifiers, the relationship between R&D intensity and performance was not statistically significant but was positive and approached significance. Thus, as firms become more diversified from their core business (up to some point), there seems to be a significant loss in efficiency of R&D activity. These firms are receiving negative returns for their investment in R&D. The positive outcomes from R&D activities must not offset the costs, thereby leading to lower performance. This suggests that the management of R&D activity becomes more complex as firms become more diversified. The results suggest that firms moving from dominant to related diversification no longer were able to manage their R&D activities effectively. The results do not support *Kelly's* (1970) contention that related diversified firms receive the greatest benefits from R&D investments.

Additionally, there were no statistically significant relationships between R&D intensity and performance for unrelated diversified firms. The results do not support *Nelson's* (1959) and *Hambrick* and *MacMillan's* (1985) arguments that diversified firms gain greater benefits from R&D by better utilizing unintended consequences. However, *Hambrick* and *MacMillan* studied R&D at the business unit rather than the corporate level. Explanation for these results may exist in the coordination costs required in diversified firms. Unrelated diversified firms have less reason to coordinate the activities and outcomes of R&D efforts than do related diversified firms. In related diversified firms, coordination is necessary to ensure synergy among R&D activities occurring across separate business units.

Schmitt (1985) noted that a large percentage of R&D activity in unrelated diversified firms occurs at the business unit level. These firms may decentralize the R&D decision authority in order to reduce coordination costs. Related diversified firms may have larger amounts of centralized R&D activities and larger R&D projects cutting across separate business units. However, *Scherer* (1965) and *Mansfield* (1968) report that diseconomies of scale in R&D occur past some point in size. More directly, *Hambrick* and *MacMillan* (1985) found a negative relationship between market share and innovative efficiency. Thus, more centralization of, greater coordination among, and larger sized R&D efforts in related diversified firms, may lead to negative returns. These data suggest that related diversified firms should find more effective ways to manage their innovation activity.

Perhaps even more concern should exist because the results suggest that industrial innovation efforts did not positively influence corporate performance during the latter half of the 1970's. Specifically, average innovation rates in 1976–1979 were 19.4% (in constant dollars) which is about 2.5 times higher than the 7.4% rate in 1972–1975 (*National Science Board* 1983). The number of full time scientists and engineers employed in companies' research and development activities increased from 353,900 in 1972, to 363,800 in 1975 (a 2.8% increase), and to 437,200 in 1979 (a 20.2% increase between 1975 and 1979) (*National Science Foundation* 1984).

Interpretation of Results

These data must be viewed in perspective. It was noted earlier that during the decade of the 1970's, national R&D expenditures in the U.S. declined consistently as a percentage of GNP and the number of patents also decreased considerably during this period. The U.S. share of the world's science and technology dropped from 33% in 1967 to less than 25% in 1980 (*Research Management* 1980). Finally, scientists and engineers, as a percent of the workforce, declined by 13% between 1968 and 1978 (*Research Management* 1980).

The "crisis of innovation" and "Not-Invented-Here syndrome" may have contributed heavily to these results. *Mendel* and *Ennis* (1985) proposed that this was a transient economic misfortune. In spite of theoretical explanations, the prevailing attitude appeared to be that R&D had an uncertain future and current instability. There was an intensified feeling of general uncertainty in industry. In corporate R&D investments, minor innovations and developments were emphasized rather than the more fundamental basic research programs. Industrial allocations in basic research in 1972–1979 grew by 18.5% in constant dollars; in applied programs, the allocations increased by 35.6% (*Research Management* 1980). Thus, innovation efforts focused more on short-term applied research. The most risky R&D projects were shifted to others (such as academic facilities) exemplifying the NIH syndrome (*Abernathy, Clark* and *Kantrow* 1983). These events seem to support *Statman* and *Tyebjee's* (1984) proposition that R&D risk was not being as comprehensively evaluated during this time period as should have been the case.

Additionally, in 1979, 85.1% of R&D funds were concentrated in corporations with more than 10,000 employees. This share had increased from 83.5% in 1972. The 100 largest companies were holding 80% of R&D money in the country in 1979 (*National Science Foundation* 1979). But concentration of economic resources and R&D potential does not necessarily yield greater innovativeness. To the contrary, this concentration often facilitates large corporations' efforts to enter the marketplace with the most trouble free product. *Peters* and *Waterman* (1982) label this policy "second to the marketplace and proud of it". The privilege of the first move, with all its disadvantages (*Porter* 1985), is granted to those for whom innovations are the only way to survive. As such, small companies have been much more innovative than large ones. There is increasing evidence that market-driven strategies tend, over time, to dampen the general level of innovation in new product decisions (*Hayes* and *Abernathy* 1980).

Part of the problem may be related to how executives measure R&D success. Generally speaking, the importance of R&D is estimated in terms of the "success" of particular projects or products. However, *Cooper* (1984a) suggests that other dimensions are involved. Included among these are:

> 1) "[...] the impact or importance of the program on company sales and profits; 2) the success rate of the program, which gauges the 'track record' of the products the firm develops; and 3) [...] the relative performance, which captures the overall performance of the program relative to objectives, to competitors, and in terms of profits versus costs."

All three dimensions have been found to be related independently to different R&D strategies. *Cooper* (1984b, 1984c) found that no group of companies performed well on all three criteria simultaneously. Similarly, *Hartwick* (1984) suggested that even short run R&D failures can contribute to long run R&D successes.

Cooper (1984b) indicated that a balanced innovative strategy can increase the rate of successful innovation. Characterized by a union of technological progress, a strong market orientation and a high degree of product fit, this strategy allows firms to concentrate their innovation efforts in high potential, high growth, non-competitive markets.

It has become clear that an appropriate R&D strategy, coupled with its effective integration into a firm's overall strategy, is critical for corporate performance (*Bozeman, Crow* and *Link* 1984, *Mendel* and *Ennis* 1985).

Implications for Managers and Researchers

Managers

The results of this study have important implications for executives because effective management of R&D investments and innovation is critical for long run firm effectiveness.

It seems that as firms become more diversified the complexity and difficulty of managing R&D efficiently increases. Thus, managers must find means to manage R&D operations in diversified businesses successfully. This is particularly true for related diversified firms. These diversified firms must take advantage of the relatedness among their various businesses to achieve high performance. As such, coordination between and control of the various businesses are necessary. Centralized control of R&D may disallow a firm's ability to serve some of the unique needs dictated by specific markets. Additionally, coordination requires forms of lateral linkages (structure) to be established. These lateral linkages could take the form of task forces or teams to share ideas and projects. Control and coordination devices, however, have direct and indirect costs. Managers of related diversified firms must develop efficient means of coordination and control of R&D activities.

In unrelated firms, no relationship between R&D intensity and performance was found. This suggests that innovation is best accomplished in these firms at the business unit level. There are few reasons or incentives for centralized control or coordination between R&D units among separate businesses. However, the results also suggest that management of R&D in the business units is uneven (some positive and some negative), balancing to produce a neutral effect. Therefore, corporate executives in unrelated firms should carefully evaluate the management practices and outcomes of each R&D unit.

Size seems to be a critical issue teaming with the level of diversification. Larger scale firms and larger R&D projects are difficult to manage effectively. Either R&D must be broken into manageable units or more effective means of managing R&D must be found in large firms. This issue is a critical one. Small firms tend to be more innovative but R&D is concentrated in large firms. Market driven R&D strategies tend to dominate in large firms. Perhaps then, large firms should develop entrepreneurial teams charged with using R&D resources effectively.

Some of the literature (research and popular) suggests that U.S. managers have been afraid of the risks associated with R&D. Perhaps the strong movement toward diversification by acquisition is partially the result of executive goals to buffer against the lack of innovation. Similarly, executives with backgrounds in finance, accounting, or law are more comfortable with acquisitions for diversification than attempts to produce innovation and growth with existing related units (*Song* 1982). In either case, this strategy can only be effective in the short run. R&D activities that result in innovation will be required in most industries to compete in either domestic or global markets. Thus, managers should become more entrepreneurial, willing to take some measured risks and seek innovation.

Finally, managers seeking innovation from R&D should select criteria or measures of R&D effectiveness carefully. The measures used should be several in number and broad in scope. This will allow innovative projects a greater probability to reach fruition.

Researchers

More research is necessary to define the exact effect of size on R&D and the innovation process. All firms in the current study were large. Differences in the innovation process between large and small firms should be examined.

Future research should seek to differentiate the effects of strategy, industry and diversification on both basic and applied research. The measure of R&D used in this study did not allow such an examination.

Additionally, future research should focus on means of establishing effective control and coordination of R&D activities in large diversified firms. The concentration of R&D investment in these firms and the complexity of managing R&D under these conditions makes this type of research highly critical. In the same light, better measures of long run R&D effectiveness should be designed.

Finally, more research is necessary to determine whether the trends of the late 1970's have extended into the latter part of the 1980's. However, data suggest that the NIH syndrome still exists. It was reported recently that 64% of firms surveyed at the end of 1984 were buying new ideas from external sources, while 82% of them employed outside consultants in their innovation programs (*New Product Development* 1984).

Conclusions

This study was conducted to examine an important, yet complex issue – the effect of innovation context. Previous findings suggested that several factors affect the effectiveness of firm innovation. This study focused on the effects of corporate growth strategies, the firm's principal industry type and the extent of the firm's diversification. The primary thrusts of this research were to enhance our understanding of this important issue and to attempt further interpretation of previous research findings.

An important finding was that of the negative relationship between R&D intensity and performance in related diversified firms. Additional analyses indicated that this negative relationship existed for firms that were using either a stable growth strategy or had pursued growth through *internal means* (rather than attempting to grow through the acquisition of other companies). This relationship also was found for related diversified firms competing principally in either the capital or producer goods industry settings. Thus, the results clearly suggest the difficulty of effectively utilizing R&D allocations in firms that have attempted to maintain commonalities (or relationships) among their product lines (or businesses). In most instances, these commonalities are sought through the sharing of markets served, distribution channels to be used and/or production technologies to be shared. Thus, it appears that the manner in which R&D investments were managed in related diversified firms in this sample was negatively related to their performance.

These findings are intriguing. However, the potential limitations of the variables and models used in this study should be recognized. For example, use of the measure industry type has some limitations. There is some evidence in the literature that in certain industries, strategic groups are formed (*Porter* 1980). These firm clusters choose to compete only in a given segment of an industry. Thus, competition develops that is unique to each strategic group. With this understanding, alternative interpretations of some results are possible. For example, the results suggest a negative relationship between R&D intensity for firms competing in a consumer durable goods industry and implementing a dominant business diversification strategy. Personal computer manufacturers produce a consumer durable good. However, the nature of competition for these firms likely varies from that experienced by other consumer durable goods manufacturers. For the PC manufacturers, it may be that R&D intensity is an important, competitive dimension. In this case, it would be necessary for all PC firms to emphasize R&D to remain competitive. However, the performance of these firms may not exceed those earned by other consumer durable goods manufacturers competing in industry settings where R&D expenditures are relatively unimportant. Additional research should be conducted to examine this kind of issue.

Viewed collectively, the findings reported herein have critical implications for executives and future research in the area. Clearly, more work is required to ensure positive returns for R&D investments. The competitive stature of U.S. and other Western nations' industrial firms in the world marketplace may depend on these efforts.

References

Abernathy, W. J., Clark, K. B., and *Kantrow, A. M.* (1983): "Mature" industries can be revitalized, *Research Management*, 4: 6–7.

Avard, S., Catto, V., and *Davidson, M.* (1982): Technological innovation – Key to productivity, *Research Management*, 25: 33–41.

Baysinger, B. D. (1985): *A contingent rational political analysis of corporate innovation*. Working paper, Texas A&M University.

Bettis, R. A. (1981): Performance differences in related and unrelated diversified firms, *Strategic Management Journal*, 2: 379–393.

Bowman, E. H. (1980): A risk/return paradox for strategic management. *Sloan Management Review*, 21:17–31.

Bozeman, B., Crow, M., and *Link, A.* (Eds.) (1984): *Strategic management of industrial R&D*, Lexington, MA: Lexington Books.

Branch, B. and *Gale, G.* (1983): Linking corporate stock price performance to strategy formulation, *Journal of Business Strategy*, 4: 40–50.

Business Week (1985): Not just another takeover, or is it? December 30: 48–51.

Business Week (1986a): Now, R&D is corporate America's answer to Japan Inc., June 23: 134–138.

Business Week (1986b): Corporate odd couples, July 21: 100–105.

Cooper, R. B. (1984a): How new product strategies impact on performance, *Journal of Product Innovation Management*, 1: 5–18.

Cooper, R. B. (1984b): New product strategies: What distinguishes the top performers, *Journal of Product Innovation Management*, 2: 151–166.

Cooper, R. B. (1984c): The strategy performance link in product innovation, *R&D Management*, 4: 247–259.

Dewhirst, H. D. (1981): *Strategic management of research and development*, Unpublished Working Paper, University of Tennessee.

Dickson, D. N. (1986): Self help from Japan, *Harvard Business Review*, 46: 8–20.

Donaldson, G., and *Lorsch, J. W.* (1983): *Decision making at the top*, New York: Basic Books.

Economics encyclopedia (1973–74): Guilfort, CT: Dushkin.

Fine, G. H. (1983): *Quality control and learning in productive systems*, Unpublished doctoral dissertation, Stanford University.

Grinyer, P. H., Yasai-Ardekani, M., and *Al-Bazzaz, S.* (1980): Strategy, structure, the environment and financial performance in 48 United Kingdom companies, *Academy of Management Journal*, 23: 193–220.

Hambrick, D. C., and *MacMillan, I. C.* (1985): Efficiency of product R&D in business units: The role of strategic context, *Academy of Management Journal*, 28: 527–547.

Hambrick, D. C., MacMillan, I. C., and *Barbosa, R. R.* (1983): Business unit strategy and change in the product R&D budget, *Management Science*, 29: 757–769.

Harrington, D. R. (1983): Stock prices, beta and strategic planning, *Harvard Business Review*, 61: 157–164.

Hartwick, J. M. (1984): Optimal R&D levels when firm j benefits from firm i's inventive activity, *Economics Letters*, 16: 164–170.

Hayes, R. H., and *Abernathy, W. J.* (1980): Managing our way to economic decline, *Harvard Business Review*, 58: 67–77.

Hitt, M. A., and *Ireland, R. D.* (1985): Corporate distinctive competence, strategy, industry and performance, *Strategic Management Journal*, 6: 273–293.

Hitt, M. A., Ireland, R. D., and *Palia, K. A.* (1982a): Industrial firms' grand strategy and functional importance: Moderating effects of technology and uncertainty, *Academy of Management Journal*, 25: 265–298.

Hitt, M. A., Ireland, R. D., and *Stadter, G.* (1982b): Functional importance and company performance: Moderating effects of grand strategy and industry type, *Strategic Management Journal*, 3: 315–330.

Hull, F., Hage, J., and *Azumi, K.* (1985): Strategies for innovation and productivity in Japan and America, *Technovation*, 2: 121–139.

Iyer, E. S., and *Ramaprasad, A.* (1984): Strategic postures toward innovation, *IEEE Transactions on Engineering Management*, EM-31: 87–90.

Kelly, T. M. (1970): *The influence of firm size and market structure on the research efforts of large multiple product firms*, Unpublished doctoral dissertation, Oklahoma State University.

Khandwalla, P. N. (1977): *The design of organizations*, New York: Harcourt, Brace, Jovanovich.

Kimmerly, W. C. (1983): R&D strategy planning in turbulent environments, *Managerial Planning*, 31: 8–13.

King, B. F. (1966): Market and industry factors in stock price behavior, *Journal of Business*, 39: 139–190.

Langetieg, T. C. (1978): An application of a three-factor performance index to measure stockholder gains from merger, *Journal of Financial Economics*, 6: 365–383.

Levine, S. J., and *Yalowitz, M. S.* (1983): Managing technology: The key to successful business growth, *Management Review*, 72: 44–48.

Linn, R. A. (1983): A sectorial approach to strategic planning for R&D, *Research Management*, 26: 33–40.

Lubatkin, M. (1983): Mergers and the performance of the acquiring firm, *Academy of Management Review*, 8: 218–225.

MacMillan, I. C., McCaffery, M. L., and *Van Wijk, G.* (1985): Competitors' responses to easily imitated new products: Exploring commercial banking product innovations, *Strategic Management Journal*, 6: 75–86.

McGinnis, M. A., and *Ackelsberg, M. R.* (1983): Effective innovation management: Missing link in strategic planning? *Journal of Business Strategy*, 4: 59–66.

Mansfield, E. (1968): *Economics of technological change*, New York: Norton.

Mendel, S., and *Ennis, D. M.* (1985): Looking at innovation strategies, *Research Management*, 3: 33–40.

Miller, D., and *Friesen, P. H.* (1986): Porter's (1980) generic strategies and performances: An empirical examination with American data, *Organization Studies*, 7: 37–55.

National Science Board (1983): University industry research relationships, Washington: GPO.

National Science Foundation (1979): An analysis of the NSF's innovation centers experiment: An effort to promote technological innovation and entrepreneurship in academic settings, Washington: GPO.

National Science Foundation (1984): Patterns of science and technology resources, Washington: National Science Foundation, 84–311.

National Science Foundation (1984): Patterns of science and technology resources, Washington: National Science Foundation, 84–311.

Nelson, R. R. (1959): The simple economics of basic scientific research, *Journal of Political Economy*, 67: 297–306.

New Product Development (1984): 4, No. 12.

Ohmae, K. (1985): *Triad power: The coming shape of global competition*, New York: Free Press.

Peters, T. J., and *Waterman, R. H.* (1982): *In search of excellence*, New York: Harper and Row.

Porter, M. E. (1985): Technology and competitive advantage, *The Journal of Business Strategy*, 3: 60–78.

Porter, M. E. (1980): *Competitive strategy*, New York: Free Press.

Rappaport, A. (1978): Executive incentives vs. corporate growth, *Harvard Business Review*, 56: 81–88.

Reilly, F. K. (1979): *Investment analysis and portfolio management*, Hinsdale, IL: Dryden.

Research Management (1980a): R&D funding in U.S. to reach $ 61.8 billion in 1980, 2: 2.

Research Management, (1980b): U.S. share of world's R&D declines, 3: 2.

Scherer, F. M. (1965): Firm size, market structure, opportunity, and the output of patented innovations, *American Economic Review*, 55: 1097–1125.

Schmitt, R. W. (1985): Successful corporate R&D, *Harvard Business Review*, 63: 124–128.

Schoeffler, S., *Buzzell, R. D.*, and *Heany, D. F.* (1974): Impact of strategic planning on profit performance, *Harvard Business Review*, 54: 137–145.

Science Indicators 1982 (1983): Washington: NSF.

Snow, C. C., and *Hrebiniak, L. G.* (1980): Strategy, distinctive competence, and organizational performance, *Administrative Science Quarterly*, 25: 317–336.

Song, J. (1982): Diversification strategies and the experience of top executives of large firms, *Strategic Management Journal*, 3: 377–380.

Statman, M., and *Tyebjee, T. T.* (1984): The risk of investment in technological innovation, *IEEE Transactions on Engineering Management*, EM–31: 165–171.

Tornatzky, L. G., and *Klein, K. J.* (1982): Innovation characteristics and innovation adoption implementation: A meta-analysis of findings, *IEEE Transactions on Engineering Management*, EM–29: 28–45.

Treynor, J. L. (1965): How to rate the management of investment funds, *Harvard Business Review*, 45: 63–75.

Treynor, J. L. (1968): The performance of mutual funds in the period 1945–1964: Discussion, *Journal of Finance*, 23: 418–419.

Weston, J. F., and *Brigham, E. F.* (1978): *Managerial finance*, Hinsdale, IL.: Dryden.

Wrigley, L. (1970): *Divisional autonomy and diversification*, Unpublished doctoral dissertation, Harvard Business School.

Chapter 5
Innovation and Human Resource Productivity in Canada: A Comparison of "High" and "Low" Technology Industries*

Barrie Gibbs, Kevin Keen, and Rob Lucas

Abstract

Improving human resource productivity through investment in "high technology" industries is but one of two options available to Canadian economic policy makers. We suggest that the human resource potential of such investment in "low technology" industries has been seriously underestimated. The paper provides an analysis of technological change in these two sectors of Canadian industry, based on a time-series analysis of 20 years of data. The first sector (referred to as "High technology") consists of the aerospace, communications equipment, electrical machinery, plastics and advanced materials industries. The second sector (referred to as "low technology") consists of the mining, pulp and paper, steel making, and industrial chemicals industries. Analysis of the data indicates that so-called "low technology" industries provide an engine for productivity equal or superior to that of the "high technology" sector.

In this paper we confront the widely-held belief that the development of high technology industries will provide the cure to most of the difficulties currently besetting the Canadian economy. We do so on the basis of unanticipated results derived from an examination of industrial data going back over two decades. These surprising results can be summed up as follows: the productivity-related outcomes of technological change occurring in basic Canadian industries compare very favourably with the outcomes associated with so-called high technology industries, both in quantity and quality. More specifically, the paper attempts to show that variations in labour productivity associated with investment in research and development are more than matched by the productivity

* The research upon which this paper is based was supported by The Max Bell Business-Government Studies Programme, and Imperial Oil Company Limited. We would like to thank Jeff Baba, David Birenbaum, and Darren Climans for their help in gathering and analyzing the data.

outcomes derived from investment in machinery and equipment. Given the current state of research on industrial innovation, these results are counterintuitive. What follows is an attempt to account for the gap between conventional expectations concerning innovative effort and the actual data on such effort.

The analysis contains both positive and negative implications for the management of human resources in Canada. Speaking positively, there is evidence that R&D expenditures may well lead to increases in the workforce, at least in the high technology industries. But, on the negative side, both industrial sectors have experienced reductions in labour productivity when, according to our best theories about innovation, R&D should have had the opposite effect. Interestingly, the low-tech sector seems better able to minimize productivity reductions than does the high-tech sector. There is evidence that the low-tech sector is investing in labour-saving methods of production, implying the future loss of jobs in this sector. It is important to keep in mind that, in Canada, the number of employees in the low-tech sector is much greater than the number employed by the high-tech sector. One of the major human resource implications of our analysis is the possibility of persistent job-loss in the low-tech sector due to technologically induced attrition. This is indeed a serious human resource management problem if those jobs cannot be replaced at a comparable rate by the high-tech sector. The data to be presented indicate that, over the past 20 years at least, such a replacement rate is not yet a reality. Therefore, on the strength of the data analysis to follow, our major human resource recommendation is the following – Canada should avoid policies that reward high-tech firms at the expense of low-tech firms. Otherwise we risk doing serious harm to the investment in human resources that *already exists* in low-tech industries, which carries with it very negative consequences for the larger economy.

Blinded by the Light – Equating Innovation with R&D

We begin by suggesting that much of the innovative effort made in Canadian industry is being overlooked. The situation stems from difficulties inherent in any attempt to study innovation, either at the level of the organization or of the industry. Innovation taking place in Canadian industry has been systematically underestimated as a consequence of the measures currently used (*Lucas* 1986). That is because the innovative effort made by Canadian firms tends to be equated with research expenditures, defined as R&D in both government and business statistics. A major reason for such a limited focus is the difficulty in assessing accurately the amount of innovative effort represented by the actual implementation of technological change over time (see *Boylan* 1977, *Downs* and *Mohr* 1976, *Marcus* 1981). Contemporary discussions of innovation in Canadian industry continue to adopt research expenditures (R&D) as the primary measure of innovative effort (*Crookell* 1984, *Hay* 1984, *Rugman* and *McIlveen* 1984).

A variety of scholars (e.g., *Boylan* 1977, *Damanpour* and *Evan* 1984, *Downs* and *Mohr* 1976, *Kamien* and *Schwartz* 1975, *Marcus* 1981, *Nelson* and *Winter* 1977, *Tornatzky* and *Klein* 1982, *Warner* 1974) have pointed out that understanding innovation requires a complex combination of theory, conceptualization, and measurement. These scholars suggest that while innovation has been the subject of many studies, the results produced have tended to provide more questions than answers. A major reason for the current ambiguity of results is that "innovation" has been used as a summary variable in a variety of ways, e.g., in the study of engineering characteristics of specific technologies, economic theories of the firm, organizational design, and strategic decision making (*Tornatzky* and *Klein* 1982). Consequently, technological innovation has become an over-used concept which has taken on a bewildering variety of meanings.

Using a definition of innovation described solely in terms of R&D may be misleading in the Canadian context for two reasons. A very large proportion of the country's economic activity occurs in the so-called "low-tech" industries. These industries are mature, and radical changes of the type produced by R&D effort are the exception rather than the rule (*DeMelto* et al. 1980). However, as *Boylan* (1977: 106) points out, the incremental accumulation of many relatively small improvements often outweighs the effects of R&D derived inventions. Secondly, the assessment of innovative effort via R&D alone fails to adequately consider the differences between two distinct types of innovation – product innovation and process innovation. The two types of innovation have been defined by *Abernathy* and *Utterback* (1978: 40) as follows:

Product Innovation:
Radical change with product characteristics in flux. New products requiring reorientation of corporate goals and/or production facilities. The creation and evaluation of new products and/or services are commonly departmentalized as Research and Development.

Process Innovation:
Incremental production process improvements leading to a rigid, efficient production system specifically designed to produce a standardized product. The production of goods or services are normally departmentalized as design or product engineering.

Since a large proportion of Canada's manufacturing activity occurs in mature industries producing highly standardized products, much innovative effort in that sector is intended to improve labour productivity rather than to develop new products. That is, technological change focuses on process innovation rather than product innovation. Consequently, the inappropriate use of R&D statistics for estimating the totality of industrial innovation, when combined with the process-oriented nature of the Canadian economy, has led to a systematic understatement of innovative effort over the past two decades. The understatement is especially acute regarding basic industrial sectors. It is here that the myth of the superiority of "high technology" industries has been born.

The Research Issues

The foregoing discussion of research on industrial innovation illustrates the first of three issues to be addressed in this paper – conventional approaches to understanding industrial innovation have relied heavily on Research and Development data, which estimates *product innovation only*. We have suggested that this focus has tended to leave the scientific community blind to innovative effort put forth through investment in *process innovation* (through expenditures on new machinery and equipment). After a discussion of our data and methodology, we will examine a second research issue – the degree of variation in labour productivity associated with each of the different types of innovative effort. Finally, we discuss some implications of alternative measures of innovative effort for future research and practice. We focus on the unexpected results these alternative measures have provided us with, and question the results of conventional approaches to innovation more generally.

The paper, despite its all-too-obvious concern with data analysis, was inspired by some interesting relations within the data on different types of industrial innovation. The effects of innovation, in one form or another, always impact on people. Consequently, we are concerned to try to place our data on industrial change into a comprehensible human context. In this paper, that context is human resource management. In trying to do so, we found little available in the literature to guide us. However, the work of *Lawrence* and *Dyer* (1983) proved to be an exception in this regard. They suggest that people in all firms face two fundamental constraints – resource scarcity (RS) and information complexity (IC) – as they go about their work. They further suggest that moderate levels of RS and IC help to increase both individual and organizational chances for adaptability and survival. Canadian high technology firms tend to face very high levels of IC (*DeMelto* et al. 1980, *Mansfield* 1984) in that the people in these firms are often dealing with technologies that are both complex and often untried. High-tech firms appear to face comparatively low levels of RS, once they have begun operation, as both individual and institutional investors clamor for an opportunity to profit from the markets created by product innovations (*Clarke* 1974, *Freeman* 1974, *Hay* 1984, *Jewkes, Sawers* and *Stillerman* 1969). Low technology firms, on the other hand, tend to face the opposite situation. Information complexity appears lower, at least in the sense that the products being produced are, for the most part, familiar to both producers and customers. As well, low-tech industries in Canada tend to face a higher level of resource scarcity due to the high degree of competition generally found in their final consumer markets. Put slightly differently, both high- and low-tech firms (in Canada at least) tend to face extremes of RS and IC. The human resource implications of this possibility are, we believe, significant. If one accepts *Lawrence* and *Dyer's* contentions, both high- and low-tech firms need to move away from the extremes they currently occupy if they are to adapt and survive over the longer term. This implies that high and low technology industries should be moving toward the *middle* of both the RS and IC scales.

With regard to high-tech firms, a move toward more moderate Information Complexity also implies the adoption and efficient use of the kinds of standard process characteristic of low-tech firms. Conversely, the implications for low-tech firms is one of increasing Information Complexity, thus implying a need for more effort to develop new products, in a way similar to that of high-tech firms. One issue that immediately surfaces is the extent to which the *existing* human resources working within each sector can be expected to deal with such change. From a policy perspective, the issue may resolve into two interesting questions – (1) is it possible for some form of *exchange* of human resources between the firms operating in the high- and low-tech sectors to occur? (2) what kinds of re-training are required, and what types of incentives and constraints are needed, in order to address the likely changes in Information Complexity?

Resource Scarcity provides a different set of challenges. Using *Thompson's* (1967) categories, low-tech firms tend to operate in "continuous process" mode, and their resource needs (in terms of the capital required to buy the necessary continuous process machinery and equipment) tend to occur relatively *infrequently*, but in *large block* amounts. Further, because of the *known* nature of the technology involved, the need for replacement machinery is also (relatively) *predictable*. High-tech firms, on the other hand, face a Resource Scarcity situation that is diametrically opposed to that of low-tech. Again using *Thompson's* ideas, high-tech firms have resource needs that are *continuous*, in terms of a constant need for the research and development effort that generates new products. However, the actual amounts of high-tech resource needs are inherently *unpredictable*, as distinct from those of low-tech firms.

It seems clear that there are significant human resource implications for firms trying to adapt to technological change. Such adaptation also begs the question of the acquisition and/or re-training of human resources. The following analysis presents the reader with the best data we could find, in order to allow him or her to decide for themselves whether Canadian trends offer any guidance to an interpretation of their own particular situation.

Industry Data

Each type of innovative effort – product innovation and process innovation – should have its own estimator which best measures it. Product innovation is conventionally estimated by R&D expenditures. There is no conventionally accepted measure of process innovation. In constructing such a measure, we will follow a distinction made by several observers (eg., *Abernathy* and *Utterback* 1978, *Boylan* 1977, *Tornatzky* and *Klein* 1982) between technological change based on R&D, as opposed to that based on *changes embodied in purchased equipment*. This distinction suggests the potential usefulness of data on *machinery and equipment expenditures* to estimate process innovation. These data, in Canada, are referred to as Machinery and Equipment Net Fixed Capital Formation (MENFCF), which is a subset of investment expenditures tabulated by Statistics Canada under the heading of Fixed Capital Formation. MENFCF represents the acquisition of new machinery and equipment over and above that required for replacement purposes[1]. Its use in this paper was also prompted by a study done in Canada by MOSST (1978) which suggests that STATSCAN R&D data significantly understate the amount of innovation in Canadian manufacturing industries. Existing R&D data do not include the importation of "invisible R&D" from foreign countries in the form of purchases of new machinery and equipment[2]. Consequently, MENFCF is a better estimator of the innovative efforts of companies which gain access to R&D done outside their own firm or industry through the purchase of new machinery, as opposed to the use of R&D alone. While some studies of the US economy (e.g., *Scherer* 1975) show that taking embodied technological change into account does not change the ranking of innovative industries, this is not the case with Canada. In this country, mining and paper-making are prime examples of industries which rely very heavily on foreign suppliers (notably Sweden, Finland, and the US) for the machinery and equipment they use. However, there are three problems associated with using MENFCF data:

(1) separating productivity effects of new equipment from the effects derived from learning and experience with the new equipment,

(2) difficulties in identifying the productivity gains accruing to end-users of new equipment due to lower unit input prices and accelerated depreciation rates, and

(3) the unknown degree to which new machinery users are also their own suppliers of new equipment (*Postner* and *Wesa* 1983).

Fortunately, MENFCF understates the amount of embodied technological change somewhat, because MENFCF data allow for the depreciation of previous capital expenditures, some of which may well have resulted in the acquisition of new technologies. We feel that this helps to compensate for the inexact nature of the measure. For the sake of conservatism it was decided that the risk of understatement was preferable to the reverse. While estimating process innovation remains an inexact science, this paper presents an analysis based on the best available Canadian data.

In order to assess both the amount and type of innovative effort made in Canadian industry, data were gathered for eight industries, all defined at the three-digit (1978 SIC) level of aggregation. The specific industries studied are identified in Table 1.

Table 1: 1978 Standard Industrial Classification and Concentration Ratios for each Industry Type

Sector	Industry Description	SIC (1978)	Number of Establishments	CR4 (1980)	CR8 (1980)
Low Tech	Steel Making	291	55	77.9	90.1
	Pulp and Paper Manufacture	271	144	30.9	49.0
	Mining	058,059	81	80.5	96.4
	Industrial Chemicals	378	134	48.8	70.9
High Tech	Aircraft and Aerospace	321	162	75.0	84.3
	Communications Equipment	335	411	52.4	64.4
	Electrical Equipment	336	253	46.8	58.1
	Plastics and Resins	373	64	57.3	75.1

In Table 1, the first four industries are meant to be representative of Canada's basic industries, operating in the areas of steel making, pulp and paper manufacture, mining, and industrial chemicals. They have been classified into a single category based on their relatively low R&D expenditures (relative to the last four industries) over the 20 year period 1963–82. They have been labelled "low technology" industries in accordance with current usage in discussion of North American industrial innovation (e.g., *DeMelto* et al. 1980, *Hay* 1984, *Mansfield* 1968, *McMillan* 1984, *Minasian* 1969, *Pavitt* and *Wald* 1971, *Rowe* and *Boise* 1974, *Thurow* 1981) as described in the following paragraph. The second, or "high technology", group was classified using the same reasoning applied in the classification of the "low technology" group. We have studied data on four industries conventionally defined as high technology industries – aerospace, communications, electrical machinery, and advanced materials.

It is important to note that there are two complementary rationales at the root of conventional definitions of a "high technology" industry. The first rationale relies on identification with specific industry characteristics, such as involvement in computing or electronic applications, as opposed to the extraction or processing of raw materials. Consequently, classification of particular industries as high-tech has been heavily dependent on the changing history of what the attributes of a high technology industry are perceived to be at particular points in time. The second rationale is based on the presence (as opposed to the absence) of significant levels of R&D expenditure associated with the industry in question. We have used these two rationales in the classification of industries found in Table 1, although we would like to point out that we do so out of respect for useful conventions rather than a belief in their conceptual adequacy.

The data used in this study are time-series variables running from 1963 to 1982 in each of the eight industries[3]. The variables used in the analysis are: gross domestic product (GDP – see Table 6), number of production-related man-hours (PRMANHR), machinery and equipment net fixed capital formation (MENFCF), and research and development expenditures (R&D). All financial data have been adjusted for inflation using the GNP deflator based on a 1971 index of 1.00. We acknowledge the problems associated with using this deflator, especially the fact that each 3-digit industry tends to face a specific market for machinery and equipment. We have decided to use the GNP deflator, despite its inadequacies, due to the lack of comparable industry-specific deflators.

General Characteristics of the Data

We now turn to a brief discussion of the data itself. From Figure 1 (GDP for both low and high tech sectors), note the sudden halt of growth in 1971 of GDP for the low-tech sector. We suggest that the drop may well be a consequence of a neomercantilist trade policy adopted by the United States when, for the first time in the twentieth century, the value of imports into the United States exceeded that of exports. As part of that policy, the United States imposed a temporary 10% import surcharge and negotiated a devaluation of the American dollar at the expense of its major trading partners – including Canada. In contrast, there is a persistent linear trend in GDP for the high-tech sector. This trend may reflect the possibility that Canadian high-tech sector products have not been entering the U. S. market in great numbers. Note also the progressive narrowing of the very large gap in GDP for both sectors after 1971. The 1982 recession is reflected in both sectors by a noticeable drop in GDP.

Turning to Figure 2, note the dominant five-year cycle for low-tech MENFCF, whereas in comparison, the R&D expenditures are quite stable. A striking feature of these data is the massive difference in the magnitudes of the two series.

There is also evidence of a similar planning cycle in MENFCF for the high-tech sector, although it is not as well defined. Note that the growth in R&D expenditures for high-tech was replaced by stagnancy in the Trudeau era from 1967 to 1979. In contrast to a commonly held belief, the high-tech sector has not been consistently increasing its innovative effort over the past twenty years. Its efforts, like those of its low-tech counterpart, have been subject to a variety of factors. Not the least of these factors are the unpredictable changes in government fiscal and trade policies.

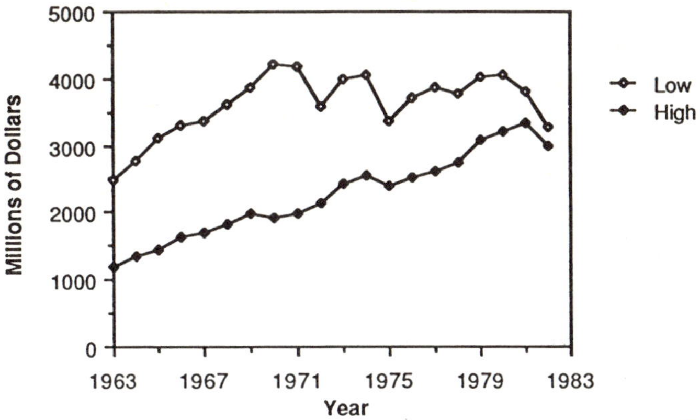

Figure 1: Gross Domestic Product for Low-Tech and High-Tech Sectors

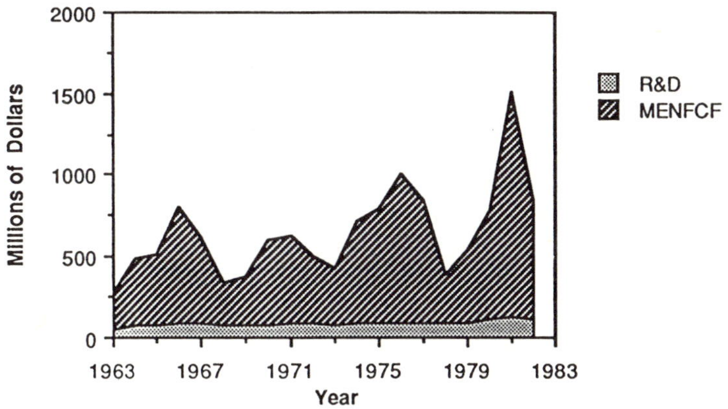

Figure 2: Research & Development expenditures and Machinery and Equipment Net Fixed Capital Formation for Low Tech

In Figure 3 it is important to note that the patterns of expenditure in R&D and MENFCF in the high- and low-tech sectors are reversed. This reflects one of the major points of this paper, namely, that these two sectors of the Canadian economy have adopted diametrically opposed approaches to technological change. We will endeavour to draw out some of the consequences of this difference in approaches to technological change and productivity improvement later in the paper.

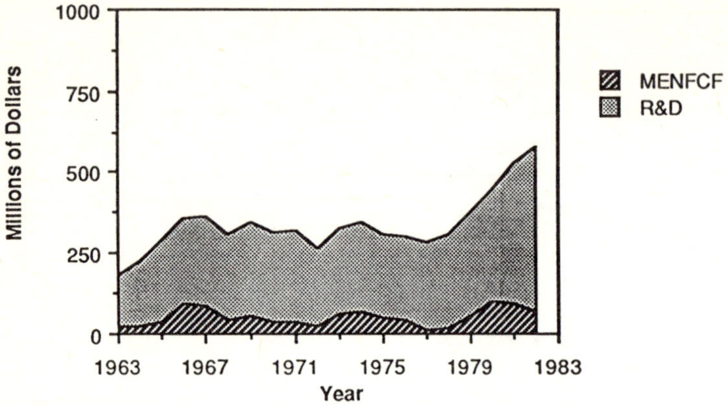

Figure 3: Research & Development expenditures and Machinery and Equipment Net Fixed Capital Formation for High Tech

Turning to Figure 4, the employment patterns in the two industrial sectors are remarkably similar, even regarding the drop in employment in 1982. However, in contrast to the GDP series presented in Figure 1, there is no progressive convergence in the PRMANHR series. One possible reason for the difference is that the low-tech sector is not replacing men with machines. It may also be that, contrary to conventional assumptions, the low-tech sector attempts to maintain highly qualified human resources while increasing productivity. These possibilities will be explored in the following sections.

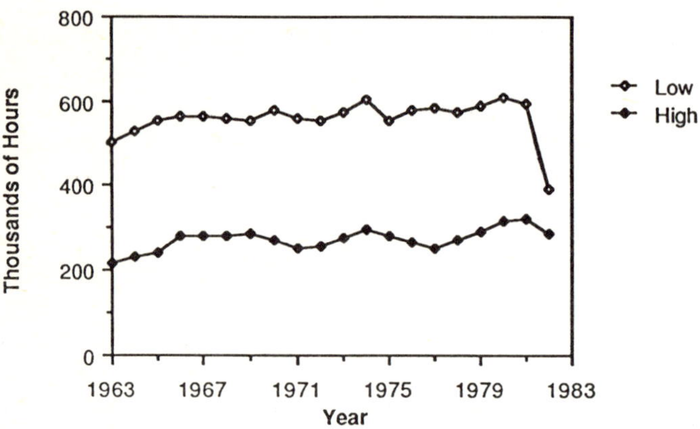

Figure 4: Production and Related Manhours for Low-Tech and High-Tech Sectors

Some Reservations Concerning the Data

If the data made available to us from Statistics Canada had resulted from a census with 100% coverage there would be no need to concern ourselves with the problem of small-sample size. The data presumably would be 100% accurate and all non-zero autocorrelations and crosscorrelations would be significantly non-zero from a statistical point of view. The task would then be to decide whether a particular autocorrelation or crosscorrelation was worthy of concern. But this would not be a statistical problem. Rather it would involve a judgement by a policy analyst. Alas, the data in our study results from a stratified survey whereby Statistics Canada samples the larger firms at a higher rate than the smaller ones. So we must contend with this sampling variability in our analysis. Moreover, we did not choose to classify all of Canadian industry as being either high-tech or low-tech. Instead, we judiciously chose several industries to be collectively representative of each classification. This is another source of variability which we must acknowledge in our statistical analysis.

There are additional limitations placed on our time-series analysis of the data by virtue of the short length of the time series being analyzed, namely, annual figures over a span of 20 years. A time series of fewer than 100 observations is considered small because the widely-used theory of time series analysis has been developed for a large number of observations, and 100 is thought to represent the lower limit in the sample size for which the asymptotic theory is applicable. There simply has not been much progress in small-sample theory for time series analysis because of the mathematical complexities of the undertaking.

In many cases, the demand for more data can be met. But it cannot be met in our analysis of innovation in Canadian industry. We wish to make an assessment of innovation which has taken place in two significant sectors of Canada's economy over the past twenty years. We are not interested in a historical survey of the past one hundred years. We do want to examine what has happened in the past 20 years, and formulate some plausible data-based explanations for current outcomes.

Statistical Methodology

The methods used to analyse the econometric data obtained for the two sectors of industry are those of multivariate time series analysis. Just as for the analysis of a single time series, there are two possible but different approaches from which to choose. One approach is an analysis in time domain. The best known example here is the Box-Jenkins method that is widely used by economists and stock-market analysts, which usually culminates in the formulation of an autoregressive moving-average (ARMA) model or some variation of the ARMA model. The other approach is analysis in frequency domain, known as spectral analysis. The strength of spectral analysis lies in its ability to detect periodicities in the time series such as the business cycle. The advantage of the Box-Jenkins method over spectral analysis is that it can be used to analyze the lag-structure of the time-series variables being examined. However, the Box-Jenkins method in time domain does do a relatively poor job of detecting periodicities in the data.

The goal of our efforts has been to assess the strength of the relationships between various input and output variables and to determine whether there exists a lag between input and output for both the low and high technology sectors of Canadian industry. This suggests that the Box-Jenkins approach in time domain is the tool to use. Nevertheless, the complementary methods of spectral analysis have been used as well. Since the results of the spectral analysis are in confirmation of the results by the time-domain approach, these results are not discussed.

Autocorrelation

The autocorrelation function is used to assess the strength of the linear relationship between values of a time series and its values in the past. Its mathematical definition is valid for any time series $L\{X(t)\}$ which is stationary up to order 2. That is, there is no trend in the series and the covariance of $X(t)$ and $X(t+s)$ is a function of the lag s alone. The autocorrelation function is defined to be

$$p(s) = \frac{\mathrm{Cov}\{X(t), X(t+s)\}}{\mathrm{Var}\{X(t)\}} .$$

When $\{X(t)\}$ is a purely random process, then the large sample distribution of $R(r)$ is Gaussian with mean O and variance N^{-1}.

The existing small-sample theory is theoretically complex and itself includes approximation. Moreover, the computations involved are not trivial. Therefore, we decided to avoid this morass and instead formulated a rule-of-thumb based on the large-sample theory. We will judge a significant autocorrelation to be present at a specified lag s if $R(r) \geq 2N^{-1/2}$. Should the large-sample theory actually apply, this rejection region would have a 5% level of significance. We, of course, do not know the actual level of significance. But this is the best we can do.

Crosscorrelation

In a vein similar to the definition of the autocorrelation function, the crosscorrelation function between two time series, $\{X(t)\}$ and $\{Y(t)\}$, is defined to be

$$R(s) = \frac{\Sigma_{t=1}^{N-|e|}[X(t+|s|) - \bar{X}][X(t) - \bar{X}]}{\Sigma_{t=1}^{N}\ [X(t) - \bar{X}]^2}\ .$$

In the nomenclature of time-series analysis, X is called the leading variable, Y is the lagged variable and s is the lag. So that this definition is meaningful, we require that both $\{X(t)\}$ and $\{Y(t)\}$ be second-order stationary, and furthermore, that $\mathrm{Cov}\{X(t)\},Y(s)\}$ be a function of $(s-t)$ only. When all three conditions are met, we say that $\{X(t),Y(t)\}$ is a stationary bivariate process, or that $\{X(t)\}$ and $\{Y(t)\}$ are jointly stationary. We estimate the crosscorrelation function at a specific lag s of the variable Y behind the leading variable X by

$$\rho_{XY}(s) = \frac{\mathrm{Cov}\{X(t),\ Y(t+s)\}}{\sqrt{\mathrm{Var}\{X(t)\}\mathrm{Var}\{Y(t)\}}}\ .$$

$$R_{XY}(s) = \frac{\Sigma_t[X(t) - \bar{X}][Y(t+s) - \bar{Y}]}{\sqrt{\Sigma_{t=1}^{N}\ [Y(t) - \bar{Y}]^2\Sigma_{t=1}^{N}\ [X(t) - \bar{X}]^2}}\ .$$

The summation goes from $t = 1$ to $(N-s)$ for $s \geq 0$, and from $t = (1-s)$ to N for $s < 0$. The expression is valid for $\{s: -(N-1) \leq s \leq (N-1)\}$.

When $\{X(t)\}$ and $\{Y(t)\}$ are uncorrelated purely random processes, the large-sample distribution of $R_{xy}(s)$ is normal with mean 0 and variance N^{-1}. This immediately suggests a rule-of-thumb whereby a crosscorrelation for which $\{Y(t)\}$ lags $\{X(t)\}$ by s time units is declared to be significant if $R_{xy}(s)\ \geq 2N^{-1/2}$. This rejection region would correspond to a 5% level of significance if all assumptions were met. In this respect our criterion is equivalent to the one chosen to seek out significant autocorrelations.

Detrending

So that an estimate can be made of the autocorrelations and the crosscorrelations of all the time series within each industrial sector, all sources of first and second order nonstationarity must be removed from offending time series. We do not believe that the data we have analyzed suffers from second order non-stationarity. But linear trend, which is a form of first order nonstationarity, is present in some of the variables we chose to examine. Linear trends can be removed by either of two methods. One method is to estimate the slope and intercept of the trend. With these estimates, one can then detrend the data and calculate the desired statistics with the detrended series. But there is penalty for making an estimate of the slope of the trend, and that is a loss of precision in the estimate of the true variance. In fact, there is a loss of exactly one degree of freedom in terms of the χ^2 distribution if the time series is Gaussian.

The other method is to perform first-order differencing by subtracting a preceeding value from the subsequent one. This will also produce a detrended time series (provided the observations are equally spaced in time). Note that this procedure is much simpler than the previous one and that is probably why it has been so popular in practice. There is, however, a loss of one degree of freedom by this latter method as well because the differenced time series has necessarily one fewer observation than the original time series.

The Gross Domestic Product (GDP) series of both low and high technology industries displays a strikingly linear trend. In comparison, there is no linear trend apparent for the input time series Machinery and Net Fixed Capital Formation (MENFCF). *Wallis* (1974) showed that the relationship between two time series is unchanged if the same adjustment is applied to both series. Moreover, he showed by way of an example that the effects on the resulting estimates may be slight. While his comments were directed towards cyclical adjustment, the setting is sufficiently general to apply to the detrending of time series by first-order differencing. Since the study of the strength of relationships among time series requires the estimation of crosscorrelations, it is absolutely essential that the trend be removed from the GDP series for both industrial sectors. This we do by first-order differencing. We follow the suggestion of *Wallis* (1974) and likewise first-order difference the remaining series: MENFCF, R&D, and PRMANHR.

Cyclical Adjustment

Business conditions never stand still. Prosperity is followed by decline. This fluctuation is referred to as "the business cycle," and it has come to characterize the economies of industrialized nations. It should be noted that some economic variables are more affected by the business cycle than others. For example, consumer paper product sales have remained remarkably stable over the past several decades, whereas sales of steel plate used in the construction of many durable consumer goods strongly exhibit the effects of the business cycle.

Although not statistically significant, the spectral analysis of each of the detrended time series of PRMANHR, MENFCF, and R&D in the low-tech sector revealed evidence of a cycle corresponding to a period of five years. This adds support to the notion of a five-year business cycle. There appeared to be evidence, although statistically insignificant, of a 6-year cycle in the low-tech GDP series but we feel that this evidence could be due to a sampling fluctuation. Similarly for the high-tech sector, the spectral analysis of the detrended time series GDP, PRMANHR, and MENFCF revealed a 6-year cycle for each, although these peaks too are not statistically significant. This particular cyclicality could perhaps be the result of a longer planning cycle in the high-tech sector. But we cannot be sure. The spectral density for R&D in the high-tech sector has a peak corresponding to a period of 19 years. We suggest the possibility that this phenomenon is a direct result of evolving changes in government policy towards investment in high-tech R&D over the period 1963–1982.

The analysis of the autocorrelation for all detrended time series of either industrial sector failed to reveal a single statistically significant autocorrelation at any lag greater than zero in agreement with results of the spectral analysis. Therefore, solid and convincing statistical evidence of any planning cycle is not available. But we believe that, given the nature of the data being analyzed, it would be imprudent to conclude that the time series are unaffected by the business cycle.

Because the business cycle is not a source of either first or second order nonstationarity, it is not essential that it be removed. Nevertheless, in some statistical analyses, it is worthwhile to correct for the presence of the business cycle. For example, most readers will be familiar with this type of correction in the generation of seasonally adjusted unemployment figures. However, we felt that in our analysis the correction was unwarranted. The work of *Wallis* (1974) indicates that we would have observed the same result in our crosscorrelation analysis even after such a correction. Since the business cycle is seen to be superimposed on each series, including series exhibiting a strong linear trend (such as GDP), it would be necessary to remove the cycle from each series. But, in the end, the final outcome would not be affected. This was the basic rationale for not removing the business cycle from the time series we have studied.

Results of Analysis – Innovation and Human Resource Productivity

Now that we have described in detail the methods used in analyzing the 20-year series of data, we turn to the major research issue of concern in this paper – the comparison of variations in human resource productivity associated with high technology versus low technology industry groups. The human resource productivity issue is discussed in five sections to follow, four of which present crosscorrelations that the analysis has shown to be of significance. We now proceed with a discussion of the first set of crosscorrelations.

For the sake of brevity in the following discussion, all series are to be taken to be the first-order differenced versions of the deflated series. The use of the first-order difference in the analysis removes the sole source of nonstationarity present in some of the series, namely, linear trend. Thus when we discuss a strong correlation at lag-zero between GDP and PRMANHR in the low-tech group, we are indeed referring to a strong correlation at a lag-zero in year-to-year change in GDP and PRMANHR. Should it become necessary to refer to the original deflated series, we will refer to it as the raw series.

The Crosscorrelations of GDP, MENFCF, PRMANHR, and R&D

With regard to gross domestic product (GDP), machinery and equipment net fixed capital formation (MENFCF), production and related manhours (PRMANHR), and research and development expenditures, the crosscorrelations judged to be significant at the 5% level and their sample values are given in Table 2.

Table 2: Crosscorrelations of Gross Domestic Product, Machinery and Equipment Net Fixed Capital Formation, Production and Related Manhours, and Research & Development Expentitures

Industry Sector	Leading Variable	Lagged Variable	Lag (year)	Crosscorrelation
Low Tech	GDP	PRMANHR	0	0.651
	MENFCF	PRMANHR	1	−0.552
			0	0.566
	MENFCF	R&D	0	0.790
	PRMANHR	R&D	0	0.532
High Tech	GDP	MENFCF	0	0.559
	GDP	PRMANHR	0	0.729
	MENFCF	PRMANHR	0	0.798

The strong correlation of PRMANHR with GDP is expected, as gains in GDP without an increase in the workforce are unlikely. With regard to high-tech industries, the correlation (at a lag of zero) between GDP and MENFCF is to be anticipated in this new and growing sector. Moreover, in the high-tech sector we see that the purchase of new equipment is closely correlated with the increase in the size of the workforce as measured by PRMANHR. This is a sure sign of growth. But, contrast this evidence with the behaviour of the well-established low-tech sector. We see that the purchase of new equipment is associated with an increase in the workforce as the new equipment is brought into production. This initial increase is followed shortly by a decrease in the workforce involved with production as the older, more labour intensive production lines are apparently shut down one year later. We can also discern the effects of an entrenched bureaucracy in the low-tech sector with R&D expenditures rising when MENFCF rises. We suggest that this effect could be a result of department heads each clamouring for their slice of the budgetary pie. The data may also represent the possibility that low-tech sector companies use their R&D groups to help put newly acquired machinery to more immediately productive use.

The Crosscorrelations of Labour Productivity, MENFCF, and R&D

There are many ways of defining labour productivity (LP). The definition that we have chosen is LP = GDP/PRMANHR. Turning to Figure 5, we see that for the high technology sector, LP has a strong positive linear trend. On the other hand, the low technology sector shows a strong positive linear trend from 1963 until a peak is reached in 1971, thereafter the series oscillates below this peak value.

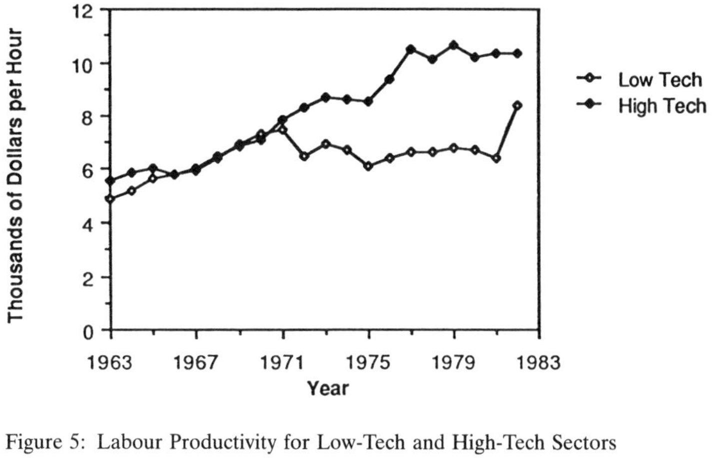

Figure 5: Labour Productivity for Low-Tech and High-Tech Sectors

Although not statistically significant, the spectral analysis of LP in both industrial sectors revealed a three-year cycle in the low-tech sector and a five-year cycle in the high-tech sector. For either level of technological sophistication, further time-series analysis requires detrending which we chose to do by first-order differencing. The crosscorrelations judged to be significant between LP and both MENFCF and R&D are given in Table 3 for both sectors of Canadian industry.

Table 3: Crosscorrelations of Machinery and Equipment Net Fixed Capital Formation and Research & Development Expenditures with Labour Productivity

Industry Sector	Leading Variable	Lagged Variable	Lag (year)	Crosscorrelation
Low Tech	MENFCF	LP	0	−0.469
			1	(0.243)
	R&D	LP	0	−0.529
			1	(0.214)
High Tech	MENFCF	LP	0	(−0.412)
			1	(−0.343)
	R&D	LP	0	(−0.266)
			1	(−0.423)

Note: Entries enclosed in parentheses are insignificant at the 5% level.

Note that there simply were no significant crosscorrelations involving LP in the high-tech sector. The general pattern of negative crosscorrelations at a lag of zero that we see for the low-tech industries was repeated, however, in the high-tech industries. The negative crosscorrelations at a lag of zero in both industrial sectors tend to confirm the possibility that firms introduce new production facilities while older existing facilities remain in production. That is, the acquisition of new machinery and equipment is recorded during the time period in which new workers are hired and trained to use the new machinery and equipment. Presumably, once the bugs have been worked out, and the workers are working near to full production potential, the old production lines are shut down and removed. The data appear to indicate that some workers are then laid off as full production is taken over by the new facilities. Therefore, one would expect to see increases in labour productivity lag behind investment in new machinery and equipment. This is indeed the case for the low-tech sector. As we see in Table 3, there is a positive correlation at a lag of one year between labour productivity and the leading variable MENFCF. The negative correlation for these two variables at the same lag in the high-tech sector is in stark contrast with this. Therefore, we are led to suspect that managers in the high-tech sector are not, for whatever reasons, phasing in new production facilities as quickly as they could.

In discussing a mixed economy such as Canada's, we ought to draw examples from both the private enterprise economy of the United States and the public enterprise system of the Soviet Union. In an example drawn from the Soviet Union, *Gorbachev* (1986) cites the case of the Bryansk Engineering Works (a producer of diesel locomotive motors) which was in the midst of a 140 million rouble retooling effort. The programme did not provide for the introduction of innovative process technologies. The number of workers had been increased by 1000 (on a base of 9000), the yield per unit of assets had dropped, and worst of all, the plant management intended to use the new production capabilities to manufacture a diesel engine of outdated design. If this example is repeated in Canada's high-tech industries, then we have grave reason to be concerned about our future as hewers of micro-chips and drawers of polymers.

Closer to home, we note that the pattern between LP and MENFCF in both high- and low-tech sectors is repeated between LP and R&D as a consequence of the high correlation between MENFCF and LP at a lag of zero. These results are in contradiction with the prevailing viewpoint in Canadian government circles that investment in R&D invariably leads to an increase in productivity (e.g., *Palda* and *Pazderka* 1982). Some productivity is necessarily lost as new production capability is added.

In addition, while Canadian unions are often blamed for adding to the problem of faltering national competitiveness (by not accepting mass layoffs on short notice as their contribution towards an internationally competitive economy), and management frequently complains about the productivity of Canadian workers, it may well be that Canadian management practices themselves ought to be more carefully examined. As well, productivity gains accruing to companies and managers as a result of periodic layoffs of skilled and unskilled workers ought also to accrue to the benefit of those workers. We would be wise to remember that it is they who, prior to losing their jobs, actually implement the otherwise abstract technological changes desired by the managerial cadre.

Designing and Evaluating an Improved Measure of Innovation

Real growth in terms of increases in GDP is only possible in the post-colonial era through innovation, population growth, or a combination of these two factors. Initially we sought to decompose innovative effort into its components of product innovation and process innovation, because industrial innovation involves either the creation of new products or the implementation of improved methods of production. But Statistics Canada does not solicit expenditure data based on these two classifications. The data are difficult to collect as the companies themselves are not generally required to maintain records of this type. Some evidence (e.g., *Lucas* 1986) suggests that innovative effort in the low technology sector is primarily concerned with *process* innovation, whereas in the high technology sector the focus is on *product* innovation. But, there are exceptions to this general observation which can be found throughout the economy. For example, the petroleum industry continues to develop new gasoline products with specialized cleaning additives, and the forest products industry has opened new mills to produce building materials in metric sizes destined only for overseas markets. We are left with two increasingly important questions – how should these innovations be classified, and what are the implications for human resource productivity?

The Potential Value of New Measures

While R&D expenditures are a good estimator of innovative effort occurring in high-tech industries, we have explained that R&D expenditures are not an appropriate measure to use in the more mature, low-tech sector. Similarly, we have explained why we expect R&D expenditures in the low-tech sector to be, in the main, related to process innovation. There is also reason to believe (e.g., MOSST 1978, *Postner* and *Wesa* 1983) that a modest base-level of R&D expenditures in low-tech firms translates into substantial expenditures on new machinery and equipment. Such R&D expenditure allows those firms to take advantage of discovered improvements in processing. In the low-tech sector, increases in R&D are correlated with increases in MENFCF at a lag of zero. As we tried to interpret the cause-and-effect aspect of this correlation, *Boylan's* (1977) review of innovation research helped by indicating that ideas on the blackboard are quite unlikely to reach the shop floor within the period of one year. It also appears quite likely that machinery and equipment purchases are co-incident with operational research done by corporate R&D groups to determine the most efficient means of utilizing newly acquired facilities. Thus it has seemed appropriate to use MENFCF as the best measure of innovation in the low-tech sector and R&D as the measure for the high-tech sector. This is, of course, what we have done up to this point in the discussion.

However, we wish to expand upon the insights drawn from a distinction which relies solely on the differences between product and process innovation. In order to do so, we suggest an alternative to the decomposition of innovative effort into process innovation and product innovation. We had adopted the process vs. product distinction because of our conviction that R&D expenditures alone are not a sufficient measure of innovation in the Canadian economy. This decision was buttressed by our belief that the emphasis of government policy on high-tech industries is detrimental to investment in what continues to be the primary engine of national human resource productivity: the country's *basic* or so-called *low-tech* industries.

But there are two difficulties to be confronted as a result of using these measures. First, if we are faced with a *middle*-technology firm, what is the appropriate measure of its innovative effort – MENFCF or R&D? Secondly, a significant number of high-tech firms operating in Canada are subsidiaries of multinational companies based in the United States and Europe. These firms import directly the fruits of R&D expenditures made by their parent firms (MOSST 1978). The problem for us here is in trying to assess the type and quantity of technology transfer within the same economic unit but between nations (see *Daly* 1987 for a more detailed discussion of this problem).

Our response to these two problems is to define a variable which can be used for both the low- and high-tech sectors, a variable which retains the properties of the relationship of MENFCF to low-tech and R&D to high-tech. Therefore, we propose the use of a new variable, *Innovative Effort* (IE), defined as follows:

IE = MENFCF + R&D

Unlike MENFCF and R&D, the Innovative Effort (IE) variable can be used for both low- and high-tech industries to assess the level of innovation. But since MENFCF dwarfs R&D in low-tech, IE behaves like MENFCF with respect to low-tech. And since R&D dwarfs MENFCF in high-tech, IE behaves like R&D with respect to high-tech. An additional benefit of the use of IE is a reduction in the number of input variables from two to one.

The Innovative Effort variable can be interpreted in another way. We know that some proportion of R&D is expended on ideas that never make their way into either the production line or the marketplace (*Palda* and *Pazderka* 1982, *Scherer* 1965). We also know that MENFCF allows for the depreciation of previous capital expenditures which have resulted in the acquisition of new technology. Therefore, MENFCF may very well underestimate process innovation. We believe that R&D data can be interpreted as a measure of *Inventive Effort*, while MENFCF can be seen as a measure of *Enterprise Effort*. Thus:

Innovative Effort = Inventive Effort + Enterprise Effort

Inventive Effort includes R&D expenditures for *both* product and process innovation. Enterprise Effort represents the sum of efforts toward implementing new processes and the enterpreneurial effort of expanding existing production to meet new markets.

The Crosscorrelations of Innovative Effort (IE), MENFCF, and R&D

Based on an analysis of the data, the relationship between IE and its constituents is given in Table 4. Only the significant crosscorrelations are reported.

Table 4: Crosscorrelations of Machinery and Equipment Net Fixed Capital Formation and Research & Development Expenditures with Innovative Effort

Industry Sector	Leading Variable	Lagged Variable	Lag (year)	Crosscorrelation
Low Tech	IE	MENFCF	0	0.999
	IE	R&D	0	0.802
High Tech	IE	MENFCF	0	0.727
	IE	R&D	0	0.813

As anticipated, we see a stronger crosscorrelation with IE and MENFCF than with IE and R&D in low-tech and the reverse in high-tech. This is a direct result of the fundamental difference in scale between R&D and MENFCF which for the low-tech sector is the reverse of that for the high-tech sector. The results of Table 4 are a natural consequence of MENFCF being the larger variable of the two measures of innovation in the low-tech sector and R&D being the larger variable in the high-tech sector.

The Crosscorrelations of Innovative Effort (IE), GDP, LP, and PRMANHR

The significant crosscorrelations among IE, GDP, LP, and PRMANHR are given in Table 5.

After using IE in place of MENFCF and R&D, the picture exhibits some subtle, but interesting, changes. In both the high- and low-tech sectors we find an increase in the labour force coupled with an increase in expenditure for innovative effort at zero lag. The largest crosscorrelations remain those between GDP and PRMANHR. There still is little evidence to support the hypothesis that increases in innovative effort produce increases in production, as measured by GDP, in either industrial sector. An increase in IE is associated with an increase in PRMANHR for the high-tech sector. While the same is true

Table 5: Crosscorrelations of Gross Domestic Product, Innovative Effort, Production and Related Manhours and Labour Productivity

Industry Sector	Leading Variable	Lagged Variable	Lag (year)	Cross-correlation
Low Tech	GDP	PRMANHR	0	0.651
	IE	LP	0	−0.474
	IE	PRMANHR	1	−0.551
	IE	PRMANHR	0	0.569
High Tech	GDP	PRAMANHR	0	0.729
	IE	LP	1	−0.499
	IE	LP	0	(−0.431)
	IE	PRMANHR	0	0.549

Note: Entries enclosed in parentheses are insignificant at the 5% level.

for low-tech, it appears that innovative effort in the low-tech sector may be devoted to reducing the labour component in production as a decrease in PRMANHR shows up in the following year. Therefore, we feel that there is some rationale for government policy encouraging investment in the high technology sector, namely, a desire to increase the number of persons employed. However, one might ask if such an emphasis ought to lead to exclusion of the employment and productivity consequences for *basic* industries. In both sectors, investment in innovation seems eventually to lead to a reduction in labour productivity. But in the high-tech sector the effects appear to be longer-lived if one overrules the possibility that the crosscorrelation of -0.431 is insignificant in practice, or at least delayed if one does not. These results, obtained by using the IE variable, were only hinted at in the insignificant results obtained in the use of MENFCF and R&D data for the high-tech sector. After using the IE variable, it seems clearer that management in the high-tech sector does not phase in new production capabilities as quickly, some might say as efficiently, as low-tech sector management. This gives us cause for concern. If we pin all our hopes for the economic future on the high-tech sector, we may find ourselves falling behind competitor nations because of poor management practices followed in the *implementation*, as opposed to the *research and development*, aspect of industrial change.

Ongoing investment and job creation in the Canadian economy may well fall off if the high-tech sector cannot produce an acceptable level of labour productivity. As well, we may find ourselves confronting an increase in labour-management strife if Canada's unionized (as well as non-union) workers are given no plausible rationale for having to accept the likelihood that their working lives will be marked by periods of unemployment and retraining implied by industrial change.

Conclusions

There are three main conclusions to be drawn from the foregoing analysis. First, the data suggest that Canadian policy-makers should re-examine the assumption that low technology industries are necessarily less innovative than their high technology counterparts. The outcomes of innovation in different industry groups may be largely determined by a distinction between product and process innovation. If this is true for Canadian industry more generally, then there are significant implications for the study and implementation of a variety of industrial policies in this country. One such policy which is of great importance concerns human resources. The analysis indicates that an industrial policy which relies very heavily on the high technology sector to create jobs needs to be re-thought for two reasons:

1. high technology industries appear to be less capable of *implementing* the fruits of R&D than are low technology industries, with significant negative consequences for human resource productivity in that sector, and
2. the adoption of process innovations by low technology industries appears to result in a drop in employment shortly after such adoption.

Therefore, it is apparent that an effective human resource policy must contain at least two elements: the encouragement of job creation in high technology industries, and the re-training of skilled workers in the low technology sector.

We suggest the usefulness of two changes in the way industrial innovation is estimated. The first change involved the use of a measure of the acquisition of new machinery and equipment (MENFCF) in order to assess process innovation. While not obviating the usefulness of R&D data, we suggest that MENFCF data allow for more meaningful conclusions regarding industrial innovation in a variety of Canadian industries, in addition to the sector usually referred to as high technology. We also defined a new variable called *Innovative Effort* which consolidates product and process innovation in Canadian industry. By using this variable, we suggest that it is considerably easier to estimate the effects of industrial innovation on important variables (such as labour productivity) across the entire range of Canadian industries.

Finally, our analysis has shown that the encouragement of R&D investment has no direct linkage with increasing GDP in either the low- or high-tech sectors. Fortunately, however, there is evidence for assuming a direct connection between R&D expenditures and increases in the workforce. Surprisingly, while the *intent of innovation* in the high- and low-tech sectors is diametrically opposite (product vs. process innovation), the *structure of innovation outcomes* in the two industrial sectors appears to be remarkably similar. Both suffer a drop in labour productivity when changes in production methods occur. But the low-tech sector seems better able to minimize the duration of the disruption than does the high-tech sector. We have provided evidence to support our contention that the low-tech sector is investing in labour-saving methods of production. That evidence also indicates the strong possibility of a net loss of jobs (over the longer run) in this sector. But there are also opportunities for a short-term increase in employment while the new production facilities are being brought on stream. We should not ignore the fact that the number of employees in the low-tech sector is much greater than the number employed by the high-tech sector. Arguably, this disparity is not likely to change in the short- to middle-term. Canada cannot afford to lose jobs due to technological attrition in the low-tech sector faster than new ones can be produced in the high-tech sector. Thus, we should avoid policies that reward high-tech firms at the expense of low-tech firms. Otherwise we risk doing serious harm to the investment in human resources that already exists in low-tech industries, which carries with it very negative consequences for the larger economy. It may seem trite to say, but if the Canadian economy is to thrive in this brave new world, federal and provincial governments must be prepared to choose policies that encourage high-tech firms to develop new products and low-tech firms to compete more effectively in traditional markets as well as developing new ones. We suggest that an important step in this direction is a more adequate understanding of industrial innovation in *both* sectors. We strongly believe that such understanding will help to increase the productivity of labour and, by implication, the international competitiveness of the Canadian economy as a whole.

References

Abernathy, W. J., and *Utterback, J. M.* (1978): Patterns of industrial innovation, *Technology Review, June/July*, 81: 40–47.

Boylan, M. (1977): The sources of technological innovations, in: *B. Gold* (ed.), *Research, technological change, and economic analysis*, Lexington MA: Lexington Books.

Clarke, T. (1974): Decision-making in technologically based organizations: A literature survey of present practice, *IEEE Transactions on Engineering Management, February* 21: 9–23.

Crookell, H. (1984): Specialization and international competitiveness, *Business Quarterly, Fall*, 49: 26–31.

Daly, D. J. (1987): Technology transfer and Canada's competitive performance, in: *R. Stern* (ed.), *Current issues in trade and investment in service industries: US-Canadian bilateral and multilateral perspectives.* Toronto: University of Toronto Press.

Damanpour, F., and *Evan, W.* (1984): Organizational innovation and performance: The problem of organizational lag, *Administrative Science Quarterly, September*, 29: 392–409.

DeMelto, D., *McMullen, K.*, and *Wills, R.* (1980): Preliminary report: Innovation and technological change in five Canadian industries, *Economic Council of Canada*, Discussion Paper No. 176.

Downs, G., and *Mohr, L.* (1976): Conceptual issues in the study of innovation, *Administrative Science Quarterly, December*, 20: 700–714.

Freeman, C. (1974): *The economics of industrial innovation*, Baltimore, MD: Penguin.

Gorbachev, M. (1986): *Political report of the communist party of the Soviet Union Central Committee to the 27th Party Congress*, The USSR Embassy Press Office, Ottawa.

Hay, K. (1984): Can Canada sustain a high-tech industry? *Business Quarterly, Fall*, 49: 52–59.

Jewkes, J., *Sawers, D.*, and *Stillerman, R.* (1969): *The sources of invention* (2nd ed.), New York, NY: Norton.

Kamien, M., and *Schwartz, N.* (1975): Market structure and innovation: A survey, *Journal of Economic Literature, March*, 13: 1–37.

Lawrence, W., and *Dyer, D.* (1983): *Renewing American industry*, New York: Free Press.

Lucas, R. (1986): High vs. low technology: Assessing innovative effort in Canadian industry, *Canadian Journal of Administrative Sciences, June*, 3: 131–145.

Mansfield, E. (1984): Technological change: A survey of findings, submission to the *Royal Commission on the Economic Union and Development Prospects for Canada*, Ottawa.

Mansfield, E. (1968): *The economics of technological change*, New York: Norton.

Marcus, A. (1981): Policy uncertainty and technological innovation, *Academy of Management Review, 6*: 443–448.

McMillan, C. (1984): *The Japanese industrial system*, New York: de Gruyter.

Minasian, J. (1969): Research and development production functions and rates of return, *Proceedings of the American Economic Association, May*: 80–86.

(MOSST) Ministry of State for Science and Technology (1978): *Importation of invisible research and development, 1974–1976* (Background Paper No. 3, July), Ottawa.

Nelson, R., and *Winter, S.* (1977): In search of a useful theory of innovation, *Research Policy, 6*: 36–76.

Palda, K., and *Pazderka, B.* (1982): *Approaches to an international comparison of Canada's R&D Expenditures*, (Report EC 22–107/1982E), Economic Council of Canada, Ottawa.

Pavitt, K., and *Wald, S.* (1971): *The conditions for success in technological innovation*, Paris: OECD.

Postner, H., and *Wesa, L.* (1983): *Canadian productivity growth: An alternative (input-output) analysis*, (Report EC 22–115/1983E), Economic Council of Canada, Ottawa.

Rowe, L., and *Boise, W.* (1974): Organizational innovation: Current research and evolving concepts, *Public Administration Review, May/June*, 34: 284–293.

Rugman, A., and *McIlveen, J.* (1984): The strategic management of Canada's multinationals: Who needs high-tech? *Business Quarterly, Fall*, 49: 64–70.

Scherer, F. M. (1965): Firm size, market structure, opportunity, and the output of patented inventions, *American Economic Review, December*, 55: 1097–1125.

Scherer, F. (1975): *The economics of multi-plant operation: An international comparisons study*, Cambridge, MA: Harvard University Press.

Thompson, J. D. (1967): *Organizations in action*, New York: McGraw-Hill.

Thurow, L. (1981): *The zero-sum society: Distribution and the possibilities for economic change*, Cambridge, MA: MIT Press.

Tornatzky, L. G., and *Klein, K. J.* (1982): Innovation characteristics and innovation adoption: A meta-analysis of findings, *IEEE Transactions on Engineering Management, February*, 29: 28–45.

Wallis, K. (1974): Seasonal adjustment and relations between variables, *Journal of the American Statistical Association, 69, (March)*: 18–31.

Warner, K. (1974): The need for some innovative concepts of innovation: An examination of research on the diffusion of innovations, *Policy Sciences, 5*: 433–451.

Appendix

A brief explanation of the abbreviations and acronyms used throughout this paper is provided for the reader's convenience in Table 6.

Table 6: List of Abbreviations and Acronyms

Abbreviations	Meaning
CR4	Concentration Ratio: percent of output of top 4 firms in an industry.
CR8	Concentration Ratio: by the top 8 firms.
GDP	Gross Domestic Product. A measure of the *output* of an economy, as opposed to a measure of the *income* (eg., GNP) recorded for all of the participants (domestic and foreign) in that economy
IE	Innovative Effort.
LP	Labour Productivity.
MENFCF	Machinery and Equipment Net Fixed Capital Formation.
PRMANHR	Production and Related Manhours.
R&D	Research and Development expenditures.
SIC	Standard Industrial Classification.
STATSCAN	Statistics Canada. A Canadian government department which is responsible for the gathering and dissemination of social and economic data.

Author Notes

1 These statistics are found in three publications: 61–205 Public and Private Investment in Canada; 61–214 Investment Statistics: Manufacturing Sub-Industries and Selected Energy Industries; 61–518 Investment Statistics: Manufacturing Sub-Industries. Fixed capital formation data from these sources is listed in *gross* amounts. Further, fixed capital formation is broken down into several components – (A) Capital expenditures, made up of 'machinery and equipment' and 'construction,' and (B) Repair Expenditures, made up of 'machinery and equipment' and 'construction.' The gross amounts of funds applied are reduced to *net* expenditures in the STATSCAN classification system by subtracting depreciation reported in any year from gross amounts reported in any year. Only those MENFCF expenditures reported under '(A) Capital Expenditures' shown above were included in the database analyzed. 'Construction' and 'Repair' expenditures were excluded from the analysis.

2 A Ministry of State for Science and Technology (MOSST 1978) report indicates two sources of "invisible R&D": (a) Canadian-owned companies which import machinery from abroad (b) Canadian subsidiaries of foreign-based multinationals which import R&D embodied in machinery and equipment acquired from the parent.

3 The sources of all data used in this paper are as follows:

Statistics Canada	Canadian Science Indicators 1983	88–201
Statistics Canada	CANSIM database	
Statistics Canada	Industrial Organization and Concentration in Manufacturing, Mining, and Logging Industries	31–402
Statistics Canada	Industrial Research and Development Statistics	88–202
Statistics Canada	Investment Statistics: Manufacturing Sub-Industries and Selected Energy Industries	61–214
Statistics Canada	Investment Statistics: Manufacturing Sub-Industries	61–518
Statistics Canada	Public and Private Investment in Canada	61–205
Statistics Canada	R&D Expenditures 1963–82	13–403
Statistics Canada	Standard Industrial Classification Manual	12–501

Section II: Human Resources and Technological Innovation

Urs E. Gattiker and Laurie Larwood

Introduction

Who uses the recent technological innovations available to business, and why? This group of four chapters represents a "micro" or person level approach to the problems encountered in the successful introduction of computers and other business technology. Two of the chapters concentrate specifically on managers' use of electronic workstations and computer technology. A third compares managerial acceptance of technology with that of support personnel in office settings, while the fourth paper looks at computer effects upon supervisory work on the manufacturing shop floor. The final chapter develops a framework for teaching employees to use new technology.

Steven Floyd, the first author in this section, investigates the use of electronic workstations by examining users' backgrounds, the content of their work, and the manner in which managers use the equipment. This comprehensive analysis measured actual use of the workstations with an on site electronic monitoring system. Several causal models are tested to determine which factors (e.g., user background, system/work fit, and skill fit) lead to high or low use of electronic workstations in managerial work.

Chapter 7 by *Jo Ann Verdin* looks at the effects of change in information technology on human resource managers. She demonstrates that they feel more aided by computer-based technology and information systems in their productivity and output quality than other managers do. The author also presents possible explanations for these rather surprising results.

Is there a relationship between acceptance of technology and career success? The third chapter in this section (chapter 8) was written by *Urs Gattiker* on the basis of research conducted with Canadian office employees. Measuring situational differences in computer attitudes based on respondents' hierarchical level and gender, the author found that people's beliefs about computers relate to perceptions of their job success, and that women perceive computers less positively than men do.

Chapter 9, written by *John Ettlie*, discusses both the evolving role of supervision in manufacturing plants as a function of new processing technologies, and the work structures and strategies used to deploy computer-based technology. Among his key results. Ettlie finds that first-line supervisors may have to deal with different levels of stress than operators, depending on the amount of routine inherent in their tasks.

Chapter 6
A Micro Level Model of Information Technology Use by Managers*

Steven W. Floyd

Abstract

This study explored three hypothesized causes of managerial use of information technology (IT) at the micro (individual) level of analysis. Exogenous variables were: (1) system/work fit, (2) system/skill fit, and (3) user background. Data from an interview were collected in the field. Use of an interactive computer terminal was the dependent variable and was measured electronically. Covariance structure analysis resulted in the respecification of two modified models. Results suggest that system/work fit is an important factor in managerial IT use. User background was not a significant factor in IT use. A revised measurement model of system use is recommended for future research.

Computer-based information technologies represent one of the most important innovations of the last twenty years. Electronic workstations (EWs) increasingly are the medium through which workers of all kinds accomplish tasks. Managers have enjoyed the benefits of information technology (IT) for some time, but until recently, their interaction with the computer has been through intermediaries (secretaries, clerks, staff analysts). Now, the picture of a manager personally using IT is a popular image. The computer on (soap opera star) "J. R." Ewing's credenza reflects society's belief that managers use EWs.

Reality is not quite consistent with the popular imagination. Only a few companies have made the desktop computer a standard piece of managerial office furnishing. There is a consensus among information systems (IS) professionals that diffusion of the innovation among managers is moving slowly (*Kirchner* 1983, *Nulty* 1984). The reluctance is surprising, since potential benefits are reported to be time savings ranging from nine to twenty-five percent (*Poppel* 1982, *Shindler* 1983). Many systems that have been installed fail for lack of *use* (*Lucas* 1981).

* The author wishes to thank Douglas Durand and James Nelson for their contributions to the dissertation upon which this paper is based. Co-editors Urs Gattiker and Laurie Larwood provided many helpful comments on the manuscript.

A number of reasons have been given for managers' disinclination to participate directly in the desktop revolution. Even before the computer invaded managers' personal workspace, *Argyris* (1971) and others found that computer-based information systems threatened managers. Use of computer output requires new skills, and computer illiterate managers find their power base eroded.

Now that managerial computer use means direct, "hands on" interaction, the list of reasons to avoid using the computer has grown. Some of the common explanations are:

(1) *Age*. Older managers are said to have a greater interest in preserving the status quo. Senior managers are unwilling to invest the time it takes to master the computer, since the impact on their remaining career will be slight.

(2) *User friendliness and ergonomics*. The technical complexity of software and interface barriers between the person and machine (i.e., the keyboard) present hurdles to managers. For many, typing is a skill that is associated with clerical work. Innovations like the "mouse" and the "touch screen" along with easy-to-use, menu-driven software are evidence of the computer industry's response to this concern.

(3) *Training and education*. Managers are frustrated as they struggle through manuals packed with jargon. Effective training sessions and tutorials are needed to make managers computer literate. Many companies offer training programs to managers. Time constraints make it difficult for them to master the technology, however.

(4) *Resistance to change*. Since adopting the technology requires personal effort and may involve embarassment, managers consciously or unconsciously resist the change process. Psychological resistance is rooted in fears about personal obsolescence and job loss.

Durand and *Floyd* (1984) explored these and other factors. They found that the perceived inability of the EW technology to support important work was paramount in explaining low usage among 54 managers. The purpose of the present study was to explore this variable, along with user background and computer skill, as causes of managerial EW use. A model is proposed and analyzed with data from interviews and electronic monitoring.

Theory

Organizing a Diverse Literature

Electronic workstations are one of the more obvious manifestations of information technology (IT) in work organizations. EWs are often considered part of a larger piece of IT called office automation (OA). The term "EW" includes terminals connected to mainframe or mini computers as well as personal computers. The functionality available to a manager through an EW varies but is some combination of the following: data communications, electronic filing and retrieval, word/text processing, electronic mail, voice record messages, interactive graphics, teleconferencing, data processing and manipulation, dictation, voice communications, and links to MIS, DSS, and expert systems.

There is a lack of integration and cumulative experience reported in the literature related to managerial IT use. Much of the material is speculative or based on the experience of a single observer. Statistically valid generalizations are rare. *Hamilton* and *Ives* (1982) reviewed articles published in the MIS field and found only 30 percent to be empirical. Forty percent of these were case studies. Only eighteen percent were field studies, lab studies, or field tests.

The relevant literature is also diverse. Studies arise from very different theoretical perspectives. One way to organize this work is to sort the ideas into eight clusters of variables. (See *Marcus* [1983] or *Olson* and *Lucas* [1982] for alternative organizing schemes.) The clusters are labeled as characteristics of: (1) the electronic (or technical) system, (2) the user as an individual, (3) the user's interpersonal relationships with co-workers, (4) the organization in which the system resides, (5) the work itself, (6) the information systems (IS) staff organization, (7) the organization's strategy, and (8) the system design and implementation process. Some research includes ideas from more than one of these categories. For example, studies of the implementation process often integrate ideas about factors affecting computer use.

This sorting of the literature suggests distinguishing "within systems" from "between systems" variation. "Within systems" variation means that the values of variables are expected to change significantly within a *single* technical system environment, i.e., at the micro or individual level of analysis. "Between systems" variation means that the values of variables are expected to change significantly between or among *multiple* system environments.

The implications of this distinction for research design are straightforward. Between system approaches are suited most for studying the effects of variables that vary across system environments: the technical system, the organization, the IS department, the design and implementation process, and the organization's strategy (e.g., *Blau* et al. 1976, *Carter* 1984, *Fuerst* and *Cheney* 1982). Within systems approaches are suited best to research about variables that vary within a single system environment: the individual, the work, and interpersonal relationships (e.g., *Ginzberg* 1981, *Kaiser* and *Srinivasan* 1982). When the system and the organization extend over the same group of people, within system studies control for technical, organizational, IS department, implementation process, and strategic variables. Consequently, the effects of individual, interpersonal, and work differences can be isolated.

Within systems variables are especially important in organizations that have already adopted the technology but face low usage among managers. In such cases, choices about the technical system, organization, strategy, and IS staff have been made. Practically speaking, these influences on managerial use may not be variable in the short run. The search for "action levers" on usage can be constrained to within systems variables (i.e. the individual, the work, and interpersonal relationships.)

In fact, EWs for managers are being implemented in many organizations with little formal analysis, planning, or design, and many implementors face the problem of managerial use *after* the equipment has been installed. This situation occurs because the managerial EW is considered an inevitable extension of an on-going office automation effort. The transfer of this "general purpose tool" from clerical and technical office workers to managers is taken for granted. In addition, individuals or groups sometimes acquire technology (e.g., micro computers) on their own. Perhaps as a result of this well intentioned but haphazard implementation, usage is often very low or unproductive, and benefits fail to materialize. Under these circumstances, the organization faces a set of questions about the causes of managerial use narrowed by the context of a given system.

This study fits into existing literature as: (1) a study conducted about variables within a single system environment; (2) a study focused on individual and work differences; and (3) a study designed to explore quantitative data.

The User's Work

In many cases of technological innovation, the design of the technology is driven by detailed analysis and knowledge of the work to be done. The fit between the technology and the work is reasonably assured by careful engineering. Problems of adoption (and thus research issues) usually are related to behavioral, psychological, or organizational factors. The case of managerial IT is different, because the question of fit between the technology and managerial work remains open (and is the subject of considerable research).

Theoretical descriptions of managerial work include: (1) lists of functions (planning, organizing, directing, staffing, coordinating, controlling, etc.; see *Barnard* 1938, *Fayol* 1949, and *Miner* 1978), (2) lists of roles (interpersonal, informational, and decisional; see *Kurke* and *Aldrich* 1983, *Mintzberg* 1973); and (3) lists of basic activities (*Kotter* 1982). These works are only a fraction of a sizable and diverse research stream that cannot be adequately reviewed here. (See *McCall, Morrison,* and *Hannan* [1978] for an excellent review of diary and observational studies.)

The conclusion common to nearly all the empirical studies of managerial work is that managers confront widely diverse demands on their time and engage in work that is fragmented, fast-paced, and interpersonally intensive.

The actual content of the work performed is described only at a general level (i.e., functions, roles, activities). While it is clear that much of what managers do involves information, it is not clear that personal use of information technology (i.e., the EW) fits the content of managerial jobs.

In the context of OA, there are studies devoted to identifying the potential work benefits from managerial EW use. Time studies of managerial activity divide the workday into meetings, document creation, reading, analysis, travel, etc. The time savings potential of managerial EW use is based on the percentage that each of these activities constitutes of the sample's time and the improvements possible in each category using EWs. Total time savings estimates vary from nine to 25 percent (*Poppel* 1982, *Shindler* 1983) as mentioned above.

Other researchers have focused on the relationship between managerial functions (or processes) and computer use. Improved decision making (*Keen* 1982, *Wagner* 1982) and communications (*Bair* 1978, *Hiltz* and *Turoff* 1979, *Moody* 1983) are often cited. *Whisler* (1970) examined the relationship between computer systems and managerial work and found that computers enhanced managerial computations and communications but had little effect on goal setting or "pattern recognition" (an aspect of managerial work unique to *Whisler's* theory). Recently, *Verdin* (1987) examined the impact of office automation and robotics on managerial jobs. She cited many specific impacts related to planning, organizing, staffing/rewarding, and controlling.

Another approach for fitting together computer systems and work comes from the socio-technical systems tradition. *Mumford* and *Weir* (1979) propose a procedure for redesigning office jobs in conjunction with implementing computer technologies. Their primary aim is to avoid the over-specialization and routinization often associated with automating clerical work. More recently, *Pava* (1983) has developed a socio-technical approach to work analysis for the nonroutine office work of managers and professionals.

In a study exploring a wide range of variables, *Durand* and *Floyd* (1984) found that the perceived fit between EWs and managerial work explained much of managers' use behavior. Associations between measures of the EW's relevance to managerial work and EW use were greater than for variables often considered crucial to managerial adoption, including: user system skills, user involvement in system design and implementation, user experience and level, and on-going user support.

Individual Differences and Managerial Computer Use

Zmud (1979) reviewed the literature of individual differences affecting the use of MIS. He sorted the research into four categories: (1) cognitive style, (2) personality, (3) attitudes, and (4) background variables. Of these four, user background variables have received considerable attention in the context of EW use. One result has been widespread acceptance among IS professionals and the general public of the idea that older, more senior managers are reluctant to use computers, especially EWs (*Buss* 1982).

The rationale for the proposition that variables such as age, managerial experience, and organization tenure affect computer use is that senior managers fear change because they have a vested interest in the status quo. Fears may be about new skills (*Tapscott* 1981), loss of control (*Galitz* and *Cirillo* 1983), or inadequacy in the face of new performance expectations (*Argyris* 1971, *Zuboff* 1982).

In his review, *Zmud* (1979) noted that general education was a background variable negatively associated with MIS, but more specific knowledge about computer systems (i.e., computer "literacy") was positively associated with MIS use. *Schultz, Ginzberg,* and *Lucas* (1984) also point to system knowledge as a predictor of system acceptance and use.

The importance of such knowledge varies with the ease of use of the computer system. *Dickson, Senn,* and *Chervany* (1977) found that complex or hard to use systems may have little impact on managers because this group usually has little knowledge about computer systems. *Couger* (1983) argues, however, that successful managerial users must be computer literate. Other researchers have pointed out the tradeoffs between ease of use and computer literacy (*Bailey* 1982, *Shneiderman* 1983).

In sum, existing theory and research point to three factors that may explain managerial EW use within a single system environment: user background, user knowledge about the system, and the user's work. Each of these has been studied in the context of managerial EWs, but they have not been examined in a multivariate framework. The present study will explore their role in a causal model of managerial IT use.

Research Questions

The modeling process is an attempt to address four research questions about managerial EW use in a single system environment.

Research Question 1. To what extent does the EW's fit with a manager's work (i.e. system/work fit) affect his/her use of the technology?

Research Question 2. To what extent does a manager's skill or literacy about the computer system (i.e. system/skill fit) affect his/her use of the technology?

Research Question 3. Can a manager's background (i.e. age, education, tenure, and experience) be an impediment to his/her use of the workstation? Are older, more senior executives prone to avoid use of the EW by virtue of these background characteristics?

Research Question 4. Of the three factors (i.e. system/work fit, system/skill fit, and user background), which is the most important cause of managerial EW use?

Toward a Model of Managerial EW Use

The model in Figure 1 proposes three constructs as causes of managerial EW use: System/work fit (ξ_1), system/skill fit (ξ_2), and user background (ξ_3). System/work fit is conceived as a latent variable representing the contribution of tasks performed on the system to work performance. System/skill fit is conceived as the ability to comprehend and manipulate the system. User background is conceived as the set of background characteristics that are associated with a propensity to avoid use of EW systems. Managerial EW use is the endogenous or effect variable in the model and is conceived as the amount of time a manager spends interacting with the computer through the EW.

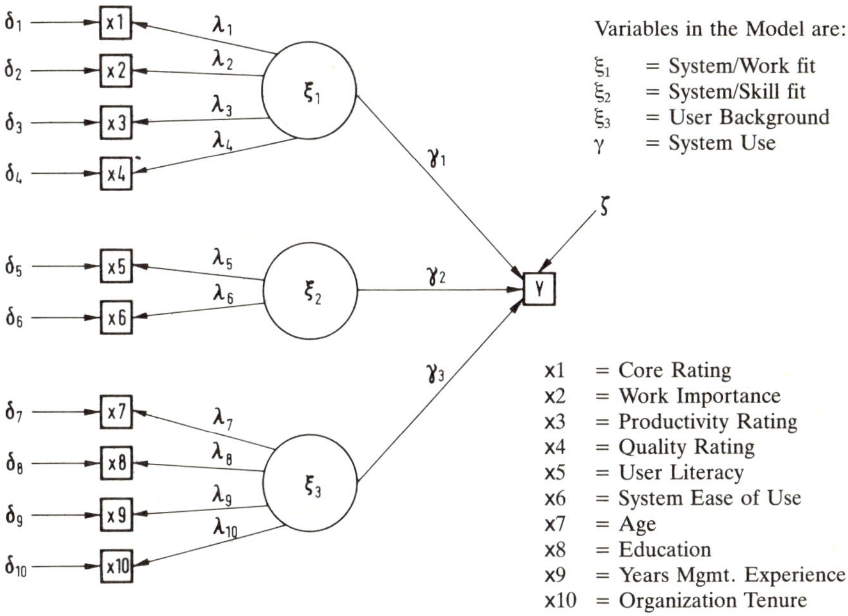

Paths in the Model are:

λ's = Parameter Estimates for the Measurement Model (validities, reliabilities)
δ's = Estimates of Unique Variance in the Measures of Exogeneous variables
γ's = Parameter Estimates for the Structural Model (estimates of effect)
ζ = The Estimate of Unique Variance in the Endogeneous Variable

Figure 1: A Causal Model of Managerial EW Use (Model 1: Initial Model)

There are measurement as well as structural relationships identified. The measurement model specifies operations that provide indicants of the underlying constructs. There are four indicants each of system/work fit and user background. There are two indicants of system/skill fit. System use (the endogenous variable) is a measured variable; it is assumed to be observable and measurable with little systematic or random error.

The model shown in Figure 1 is an initial model determined by the author's interpretation of existing research. The relationships specified imply hypotheses. Since this area of investigation is relatively new, however, there was no basis for limiting analysis to strict "hypothesis testing." The point of the analysis was to explore the model's fit with an empirical structure.

Method

Research Design

This was a "passive observational" field study (*Cook* and *Campbell* 1979). Data were collected at a single point in time, and the structure of observed covariance among variables was analyzed. Some degree of control over "between systems" variation was achieved by collecting data within a single system environment.

The Sample

The sample consisted of 110 managers working at a headquarters location for a large, multinational company. The participants were classified into three groups; there were 11 executives, 57 managers, and 42 professionals. They were spread evenly across a variety of functional areas. All had EWs in their offices that had been installed for twelve months or more.

The job classes (executives, managers, and professionals) represented the complete range of hierarchical levels in the professional ranks of employees at the study site. Executives were uppermost, middle-level "managers of managers" responsible for an entire department. Managers supervised professional staff within the department. Professionals were nontechnical analysts in business-oriented functions (personnel, accounting, finance, logistics, planning, information systems) who had no supervisory duties. Most of the professionals (35 out of 42) had substantial supervisory experience in their backgrounds.

The sample is arguably a "modal" or "target" instance (*Cook* and *Campbell* 1979) of the population to which generalizing is appropriate, i.e., managers who are experienced users of EW systems. The organization was committed to implementation and had devoted substantial resources for training, hardware, software, and support.

To find a sample of managers with experience in a computing system of substantial functionality, however, it *was* necessary to study a progressive organization. Variables that are not representative of "typical" organizations (e.g., unusually favorable attitudes toward computers) were a concern. Misgivings about unrepresentative enthusiasm for computers in this sample were allayed by: (1) the relatively modest level of EW use observed, (2) the willingness of interviewees to express negative attitudes about the system, and (3) the wide diversity in the attitudes of interviewees. Moreover, since inferences based on the model concern relationships among variables rather than mean values of variables, unusually high EW use levels in the sample are not a concern per se. (Further description of the sample is available form the author upon request.)

Measures

The study employed two data collection methods. Most of the indicants of independent variables (except X5) were measured in a structured interview. User literacy (X5) and EW use (Y) were measured by the computer system itself.

System/work fit (ξ_1). Four sets of interview items were constructed to measure system/work fit. First, participants were asked to rate each system function on a scale from zero to 100 percent to represent the proportion of use that contributed to their core work responsibilities. The functions rated were electronic mail, document preparation, data inquiry and analysis, graphics, electronic filing, electronic calendar, and organization directory. The total core rating (X1) was calculated for each participant by summing core percentages across all functions. Second, the importance of each function to the participant's work was scored on a Likert-type scale, and the overall importance of the EW to the manager's work (X2) was measured by summing these ratings. Finally, productivity (X3) and quality (X4) impacts of the system were measured on Likert-type scales ranging from very negative, neutral to very positive. After describing specific incidents of productivity/quality impact, participants were asked to rate the global impact of the system on their work using these scales.

System/skill fit (ξ_2). This construct was measured in two different ways. First, the user's facility or literacy with the EW (X5) was measured as an integer representing the number of unique commands invoked. Use of many different commands demonstrated a wide familiarity with the system's functionality. Fewer commands represented a smaller "vocabulary" and resulted in a lower literacy score. Commands were monitored electronically over at least a three week period. To measure the system's ease of use (X6), a Likert-type scale was designed to assess the amount of undesirable and/or unnecessary effort the system required of the user. The scale ranged from "no undesirable/unnecessary effort" to "excessive" effort. A high rating on this scale indicated the user's perception that the system was difficult.

The combination of these two indicants provided the measure of system/skill fit. This was consistent with the belief that the construct represented the matching of the system's ease of use with the user's ability to use.

User background (ξ_3). The background characteristics of the user measured were age (X7), education (X8), management experience (X9), and tenure in the organization (X10).

Managerial EW Use (Y). Use of the EW system was measured by electronic monitoring. Software was developed that monitored the time spent by a user within each EW system function (listed above). The computer clock started when a particular command initiated a function and stopped with a command to exit the function. Each participant was monitored for a period of three to four weeks. The data was carefully screened and reviewed by the user him/herself to control for inaccuracies created by interruptions in a session. This procedure seemed to be sufficiently reliable and face valid to justify treating EW use a measured (rather than latent) variable.

Computer monitoring of individual behavior is controversial. Under some circumstances, monitoring violates ethical principles (*Mason* 1986). In other cases, monitored users may alter their behavior (*Andrew* 1983). The use of IT to observe behavior unobtrusively serves the researcher's purpose but may raise issues about the basic purpose of IT in organizations. (See *Zuboff* 1985 for a discussion of IT's purpose.)

In this case, two factors mitigated the negative side of monitoring. First, participation in the study was voluntary, and the monitoring was initialized and terminated in the participant's presence. The study's goal was clearly stated: it was an evaluation of the system, not an evaluation of users. Second, use of the system itself was voluntary. There were no formal expectations about appropriate computer use. Thus, participants as a whole probably had no basis to alter the level or nature of their interaction with the terminal. The possibility that particular individuals modified their use pattern as a result of monitoring cannot be ruled out, however.

Analysis

Causal models have been used in formulating and testing theories in the social and behavioral sciences, though they are relatively new in management research. These models specify theory in terms of cause and effect variables as well as their indicants. Path coefficients are parameters describing causal and measurement relationships. LISREL VI provides maximum likelihood (ML) estimates of the parameters and statistics to assess the goodness-of-fit.

Anderson and *Gerbing* (1982), *Bagozzi* (1983), and others have argued for an iterative approach to causal modeling. They insist that an adequate specification of the measurement model is necessary before meaning can be assigned to the analysis of structural relationships. The lack of construct unidimensionality is a frequent problem with initial models (*Anderson* and *Gerbing* 1982), and LISREL provides several statistics that suggest means for improvement. If results are interpreted cautiously and in an exploratory spirit, refitting modified models can avoid becoming a "fishing expedition."

James, Mulaik, and *Brett* (1982) describe the model revision process as analyzing a sequence of "nested" models, i.e., "submodels" with variables contained in an initial or general model. When the status of a parameter changes between the initial and a nested model, a comparison of the two sequential models serves as a test of the importance of the parameter. To avoid too much "exploration," the revisions to the initial model should be theoretically as well as empirically defensible.

In the present study, the measurement relationships of the initial model were explored first. Then a revised or nested model, M2, was developed with more reliable exogenous latent variables (by fixing certain measurement parameters to zero) for inclusion in the structural equations. A third model (M3) was introduced to explore the treatment of the endogenous variable as a latent construct while otherwise retaining the measurement model from M2. While these iterations do draw upon post hoc information, the modifications can be defended on substantive grounds. In short, there is some "constrained exploration" in the modeling process, but it is not a "fishing expedition."

Results

Findings About the Initial Model (Model 1)

Table 1 shows the Pearson product moment correlations among the measured variables. *Long* (1983) notes the importance of casewise deletion of missing data in deriving correlations for ML analysis; this procedure had the effect of reducing the sample size to 73. Table 2 shows parameter estimates, standard errors and t-values provided by LISREL for the initial measurement model (M1). Latent variables were standardized for purposes of setting their metric.

There is evidence of misspecification in the measurement model. The estimate for λ_4 is too low relative to other measures of ξ_1. The path coefficient λ_8 is

Table 1: Correlations, Means and Standard Deviations (N=73)

	Y	X1	X2	X3	X4	X5	X6	X7	X8	X9	X10
Y	1.0										
XI	.46	1.0									
X2	.53	.94	1.0								
X3	.41	.48	.47	1.0							
X4	.20	.14	.24	.37	1.0						
X5	.85	.53	.59	.54	.33	1.0					
X6	.09	.22	.28	.15	.34	.17	1.0				
X7	−.20	−.11	−.16	−.12	−.03	−.23	.27	1.0			
X8	−.12	−.10	−.13	−.04	.01	−.17	−.15	−.26	1.0		
X9	−.16	−.14	−.23	−.06	−.07	−.21	.19	.81	−.36	1.0	
XI0	−.29	−.24	−.29	−.19	−.13	−.42	.03	.48	−.13	.59	1.0

	Mean	S. D.
Y	20.88	19.77
XI	645.16	455.94
X2	6042.15	3865.71
X3	7.14	1.18
X4	6.80	1.16
X5	14.59	9.14
X6	4.39	2.27
X7	43.87	7.37
X8	14.79	1.59
X9	18.49	7.50
XI0	8.23	7.90

negative and contrary to the *a priori* definition of the latent variable (cf. Table 2). λ_5 and λ_6 suggest that X5 and X6 are not measuring a common underlying construct. The construct reliability (*Bearden, Sharma,* and *Teel* 1982) for ξ_2 is only .33; reliabilities for ξ_1 and ξ_3 are .79 and .69, respectively. These findings suggest revising the measurement model by dropping X4 and X8 to improve the measurement of ξ_1 and ξ_3. In addition, since ξ_2 was not measured adequately, it is dropped in further analysis.

Findings About Modified Models

Model 2

Figure 2 shows the parameter estimates and standard errors for a model (M2) with the modifications suggested by the analysis of the initial measurement model. Several measures of the overall adequacy of the model are shown. Note the consistency between the measurement parameter estimates in M1 and M2.

Table 2: ML Estimates for the Initial Measurement Model (MI)

Parameter	Estimate	Standard Value	*t*-Value*
λ_1	.93	.09	10.40
λ_2	1.00	.08	11.89
λ_3	.46	.11	4.09
λ_4	.24	.12	2.07
λ_5	.61	.11	5.51
λ_6	.28	.12	2.33
λ_7	.80	.10	7.69
λ_8	−.35	.12	−3.09
λ_9	1.00	.10	10.43
λ_{10}	.58	.11	5.24
δ_1	.13	.04	3.01
δ_2	.00	.04	n/a
δ_3	.79	.13	6.00
δ_4	.94	.16	6.01
δ_5	.63	.11	5.72
δ_6	.92	.16	5.97
δ_7	.36	.09	4.12
δ_8	.87	.15	5.99
δ_9	.00	.10	n/a
δ_{10}	.66	.12	5.76

*Critical two-tailed *t*-values: $t \leq .10 = 1.282$, $t \leq .05 = 1.645$, $t \leq .005 = 2.576$.

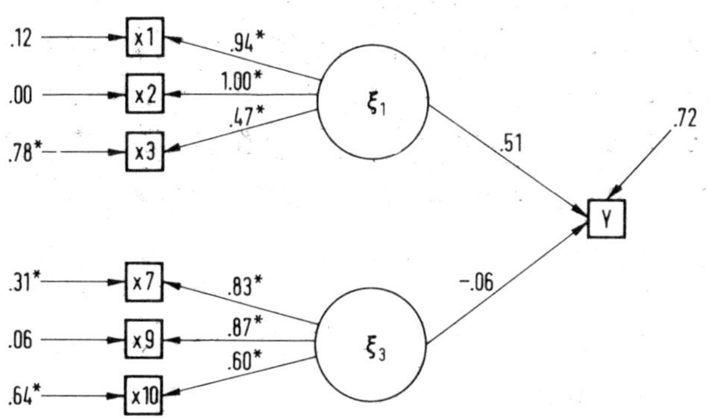

Chi-square	18.13
D.F.	13
Chi-square/D.F.	1.40
GFI	.76
Incremental Fit Index	.96
Coefficient of Determination	.28

*p ≤ .005, two-tailed

See Figure 1 for a list of variable names and path labels.

Figure 2: Model 2: (M2) Modified Model

M2 fits the data well. The chi-square value is small enough to suggest that the covariance structure presumed by M2 is not significantly different from the observed covariance structure (p = .153). The chi-square/D. F. ratio is well below the value of 10.0, identified by *Schmitt* and *Bedeian* (1982) as indicating an acceptable fit. The incremental fit index is above the value (.95) usually associated with well fitted models (*Bentler* and *Bonnett* 1980). The GFI represents the relative amount of variances and covariances jointly accounted for by the model (*Jöreskog* and *Sörbom* 1983). These statistics suggest that M2 is a good candidate for testing in future research.

Model 3

The definition of system use (Y1) as "time interacting with the system" was appealing initially for its clarity. It was narrower than that proposed by *Barkin* and *Dickson* (1977), however. They define utilization as the inclusion of system generated data by decision-makers in their Human Information Processing system (p. 36). This suggests that system use is an unobservable construct rather than a measured variable as it was treated in Model 1. When the number of system commands used (X5) is coupled with time of use (Y1), the two may be indicants of both the variety and level of use. This operationalization seemed promising enough to be included in a revised model.

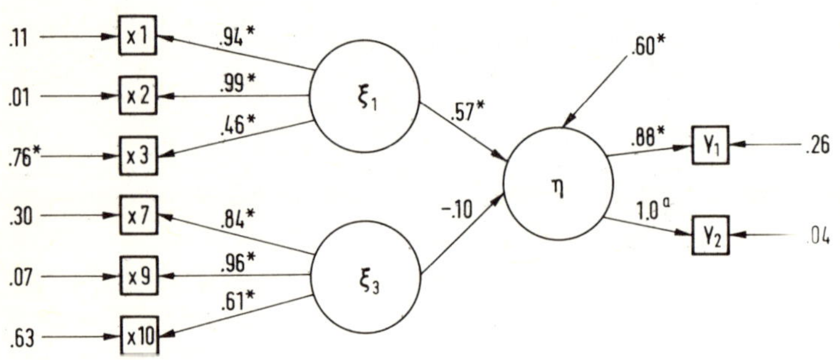

Chi-square	33.77
D.F.	18
Chi-square/D.F.	1.88
GFI	.70
Coefficient of Determination	.38

*p ≦ .005, two-tailed
a – Fixed parameter value

See Figure 1 for a list of variable names and path labels.

Figure 3: Model 3: (M 3) Modified Model

Thus, a third model (M3) was estimated with a revised measurement submodel for the dependent variable. Figure 3 shows the parameter estimates, standard errors, and fit indices. In M3, managerial EW use is treated as a latent variable indicated by both the amount of time spent with the system (Y1) and the variety of commands used (Y2).

The measurement parameters for exogenous variables and the structural parameters in M3 are consistent with those of the prior models. It is interesting to note that the fit of M3 is not quite as good as that of M2, but the multiple R-squared is significantly greater.

User background, ξ_3, plays no significant role in the structural equations of either M2 or M3. In M3, 97 percent of the explained variance in the dependent variable is accounted for in ξ_1; in M2, the figure is 98.6 percent of explained variance. User background is not a significant cause of managerial use in the model.

Discussion

The role of system/work fit in a causal model of managerial EW system use was explored for the first time in this study. Results suggest that this construct is an important cause of use. The second major conclusion is that user background is not causally related to managerial use of an EW when system/work fit is taken into account. Note that bivariate correlations between measures of user background and EW use were significant (see Table 1); this is consistent with the findings of other researchers (e.g., *Lucas* 1975, *Poppel* 1982, *Werner* 1974). When system/work fit is introduced into the equation, however, these associations become nonsignificant.

This finding highlights the importance of multivariate approaches in research about computer use. Associations between background variables and use may be accounted for by a third variable, system/work fit. In other words, older, senior managers do not use the system because the system/work fit is low – not because they cherish the status quo. In fact, older, senior managers who have discovered a fit between the system and their work are as likely to use EWs as more junior counterparts.

Results from the analysis also suggest that use of an EW means more than time spent with the terminal. The variety of commands used may be another good indicator of EW use. For example, a very sophisticated user may know the system well and spend less time on the terminal than the novice, but the time spent by the experienced user would draw on more of the system's functionality. One would want to count the use of the sophisticated user as greater than that of the novice. This reasoning also suggests that electronic measures should be coupled with subjective measures of EW system use. (*Barkin* and *Dickson*

[1977] identified one approach to the subjective measurement of information system use.) Future research should explore the unidimensionality of the use construct measured in these three ways.

Two limitations for generalizing the findings of this research should be noted. First, the fact that system/skill fit could not be introduced into the structural equation means that it may act as an unmeasured third variable that accounts for what appears to be a relationship between system/work fit and use. Also, there are other unmeasured third variables that may threaten the validity of the findings. In addition, the size and nature of the sample may bias results. Since individuals were selected from a single organization, results need validation in other settings.

The study lacks evidence about the causal directionality in the model. This is an issue for the path between system/work fit and use. (It is nonsense to suggest that EW use causes user background.) The reciprocal path was estimated using formulas from *Jöreskog* and *Sörbom* (1982) and found to be significant. Indeed, it seems likely that managers increase system/work fit as they use the system more. Future research should be designed to explore the reciprocity between system/work fit and managerial EW use.

Despite these limitations, the results of this study point to the importance of fitting the EW system to the work of managers. The stereotype of the reluctant manager appears to be unwarranted. Even senior managers will use the EW, if the tool fits the work they have to do.

Implications for Managers and Researchers

Practicing managers confronted with the wave of EW technologies face the need to justify both personal and organizational adoption. The search for managerial work benefits should go beyond time studies of routine activities (making phone calls, attending meetings, scheduling appointments, etc.) to explore the *purposes* of managerial effort. Managers work toward goals that are linked to the objectives of the organization. Thus, the place to begin looking for important managerial applications is in the goals of the organization itself.

Instead, the tendency among early adopters of managerial EWs has been to find benefits from avoiding "telephone tag," substituting teleconferences for travel, reducing time spent coordinating schedules for meetings, and so on. These benefits have the primary advantage of measurability and may even provide the basis for justifying equipment purchases. Justifying the time expended to learn and use the EW, however, requires looking beyond time saved on peripheral activities.

When systems are designed and used to contribute to the organization's mission, the real payoffs from IT use can be realized. For example, one of the departments studied was responsible for managing the logistical plans for the division. The department manager created and maintained a master "plan of plans" in a file that could be read and revised by each of his subordinates. By assuring the absolute currency of the overall plan, the department manager was able to authorize significant adjustments in a very tight reporting schedule without undermining the integrity of the plan. This use of the EW enhanced the manager's core work and contributed to his organization's purpose.

Researchers interested in the relationship between managerial work and EW use share with practitioners the tendency to focus on time savings. While it is difficult to operationalize the concept of goal contribution, theories of managerial work (e.g. *Kotter* 1982, *Miner* 1978, *Mintzberg* 1973) provide an alternative to time studies. In one sense, these theories point to the common goals of managers.

Consider, for example, the subtle use of the electronic calendar made by one executive. Prior to delegating a project, the executive reviewed the calendar of each manager to determine prior work commitments. This use of the electronic calendar saved no time in keeping up with appointments; rather, it met the information needs of delegation.

Linking EW use to the components of core managerial work is a fruitful area of inquiry for future research. Too often, researchers and practitioners alike have focused on applications that are at best peripheral to the manager's job. As the above examples illustrate, there *are* potential uses of the EW that are more central to the manager's job, and in the long run, managers may discover these on their own. In the meantime, reseachers can contribute by describing the core uses managers have already developed. When theories of managerial work are combined with this descriptive base, managers may have more reason to become full participants in the information technology revolution.

References

Anderson, J. C. and *Gerbing, D. W.* (1982): Some methods for respecifying measurement models to obtain unidimensional construct measurement, *Journal of Marketing Research*, November, 19: 453–459.

Andrew, J. (1983): As computers change the nature of work, some jobs lose savor, *The Wall Street Journal*, 206: May 6: 1–15.

Argyris, C. (1971): Management information systems: The challenge to rationality and emotionality, *Management Science*, 17 (6): B275.

Bagozzi, R. P. (1983): Issues in the application of covariance structure analysis: A further comment, *Journal of Consumer Research*, 9 (March): 449–450.

Bailey, R. W. (1982): *Human performance engineering*, Englewood Cliffs, N. J.: Prentice-Hall, Inc.

Bair, J. H. (1978): Communications in the office of the future, Proceedings of the International Computer Communications Conference, Kyoto, Japan: Stanford University Press.

Barkin, S. R. and *Dickson, G. W.* (1977): An investigation of information system utilization, *Information and Management*, 1: 35–45.

Barnard, C. I. (1938): *The functions of the executive*, Cambridge, MA: Harvard University Press.

Bearden, W. O., Sharma, S. and *Teel, J. E.* (1982): Sample size effects of chi square and other statistics Used in evaluation of causal models, *Journal of Marketing Research*, 19 (November): 425–430.

Bentler, P. M. and *Bonnett, D. G.* (1980): Significance tests and goodness of fit in the analysis of covariance structure, *Psychological Bulletin*, 88: 588–606.

Blau, P. M., Falbe, C. M., McKinley, W., and *Tracy, P.* (1976): Technology and organization in manufacturing, *Administrative Science Quarterly*, 21 (March): 20–40.

Buss, M. J. (1982): Making it electronically, *Harvard Business Review*, January-February 60 (1): 89.

Carter, N. M. (1984): Computerization as a predominate technology: Its influence on the structure of newspaper organizations, *Academy of Management Journal*, 27 (2): 247–270.

Cook, T. D. and *Campbell, D. T.* (1979): *Quasi-experimentation*, Boston: Houghton Mifflin Co.

Couger, J. D. (1983): Developments to facilitate managerial use of the computer, *Computing Newsletter*, 16 (8): 1.

Dickson, G. W., Senn, J. A., and *Chervany, N. L.* (1977): Research in management information systems: The Minnesota experiments, *Management Science*, 23 (9): 913–932.

Durand, D. E. and *Floyd, S. W.* (1984): Managerial use of office automation: A path analysis, Proceedings of the Academy of Management, Boston, MA.

Fayol, H. (1949): *General and industrial management*, translated by Constance Storrs, London: Pitman Publishing.

Fuerst, W. L. and *Cheney*, P. H. (1982): Factors affecting the perceived utilization of computer based decision support systems in the oil industry, *Decision Sciences*, 13 (4): 554.

Galitz, W. O. and *Cirillo, D. J.* (1983): The electronic office: How to make it user friendly, *Management Review*, 12, April: 24.

Ginzberg, M. J. (1981): Early diagnosis of MIS implementation failure: promising results and unanswered questions, *Management Science*, 27 (4): 459.

Hamilton, S. and *Ives, B.* (1982): MIS research strategies, *Information and Management*, 5: 339–347.

Hiltz, S. R. and *Turoff, M.* (1979): *The network nation*, Reading, MA: Addison-Wesley.

James, L. R., Mulaik, S. S., and *Brett, J. M.* (1982): *Causal analysis*, Beverly Hills, CA: Sage.

Jöreskog, K. G. and *Sörbom, D.* (1982): Recent developments in structural equation modeling, *Journal of Marketing Research*, 19 (November): 404–416.

Jöreskog, K. G. and *Sörbom, D.* (1983): *Analysis of linear structural relationships by maximum likelihood and least squares methods*, Chicago: National Educational Resources, Inc.

Kaiser, K. and *Srinivasan, A.* (1982): User-analyst differences: An empirical investigation of attitudes related to systems development, *Academy of Management Journal*, 25 (3): 630.

Keen, P. (1982): DSS balances skills of managers, machine, *Computerworld*, 16 (44): 39.

Kirchner, J. (1983): Lack of managerial support systems lamented, *Computerworld*, 7, March 21: 14.

Kotter, J. P. (1982): *The general managers*, New York: The Free Press.

Kurke, L. B. and *Aldrich, H. E.* (1983): Mintzberg was right!: A replication and extension of the nature of managerial work, *Management Science*, 29 (8): 975–984.

Long, J. S. (1983): *Confirmatory factor analysis*, Beverly Hills, CA, and London: Sage.

Lucas, H. C. (1975): *Why information systems fail*, New York: Columbia University Press.

Lucas, H. C. (1981): *Implementation: The key to successful IS*, New York: Columbia University Press.

Marcus, L. M. (1983): The new office: More than you bargained for, *Computerworld*, 17 (8a), February 23: 35.

Mason, R. O. (1986): Four ethical issues of the information age, *MIS Quarterly* 10, (1) January: 5–12.

McCall, M. W. Jr., *Morrison, A. M.*, and *Hannan, R. L.* (1978): *Studies of managerial work: Results and methods*, Technical Report No. 9, Greensboro, NC: Center for Creative Leadership.

Miner, J. B. (1978): *The management process: Theory, research and practice*, New York: MacMillan.

Mintzberg, H. (1973): *The nature of managerial work*, New York: Harper and Row.

Moody, H. G. (1983): Matching function to need, computers enhance our work, *Office*, 97 (1): 101.

Mumford, E. and *Weir, M.* (1979): *Computer systems in work design – The ETHICS method*, New York: John Wiley and Sons.

Nulty, P. (1984): How personal computers change manager's lives, *Fortune*, 112: September 3: 38–48.

Olson, M. H. and *Lucas, H. C.* (1982): The impact of office automation on the organization: Some implications for research and practice, *Communications of the ACM*, 25 (11): 838.

Pava, C. H. P. (1983): *Managing new office technology*, New York: The Free Press.

Poppel, H. L. (1982): Who needs the office of the future? *Harvard Business Review*, 60 (6): 146.

Schmitt, N. and *Bedeian, A. G.* (1982): A comparison of LISREL and two stage least square analysis of an hypothesized life-job satisfaction reciprocal relationship, *Journal of Applied Psychology*, 67: 806–817.

Schultz, R. L., *Ginzberg, M. J.*, and *Lucas, H. C.* (1984): A structural model of implementation, unpublished working paper.

Shindler, P. E. (1983): Exec work study statistics seen limited in value, *Information Systems News*, January 24: 28.

Shneiderman, B. (1983): Human factors of interactive software, *Proceedings for IBM Heidelberg German Scientific Center 15th Anniversary Symposium*, March 18, *A. Blaser* and *M. Zoeppritz* (Eds.), Springer Verlag, W. Germany.

Tapscott, D. (1981): How office automation will alter the corporation, *Executive*, 23 (7): 50.

Verdin, J. A. (August, 1986): Office automation: How it has changed managerial jobs, Paper presented at the 46th Annual Meeting of the Academy of Management, Chicago.

Wagner, G. R. (1982): DSS: Dealing with executive assumptions in the office of the future, *Managerial Planning*, 30 (5): 4.

Werner, D. J. (1974): A study of the use of an experimental information system in a medical environment, *Management Information*, 3 (3): 133–142.

Whisler, T. L. (1970): *Information technology and organizational change,* Belmont, CA: Wadsworth Publishing Co.

Zmud, R. W. (1979): Individual differences and MIS success: A review of the empirical literature, *Management Science*, 25 (19): 966.

Zuboff, S. (1982): New worlds of computer mediated work, *Harvard Business Review*, 60 (5) September-October: 142–152.

Zuboff, S. (1985): Automate/informate: The two faces of intelligent technology, *Organization Dynamics*, 14 (2), Autumn: 4–18.

Chapter 7
The Impact of Computer Technology on Human Resource Information System Users*

Jo Ann Verdin

Abstract

The research reported here examines the impact of information system utilization on attitudes about jobs, careers, decision making processes, productivity and quality of output for a sample of 107 users of human resource information systems in three organizations. In addition, the impact on human resource department users and users from other areas is compared. It was found that skill requirements and inter-departmental communications increased, career opportunities were enhanced outside the organization to a greater extent than within the firm, some improvement in the quality of decisions resulted, and productivity and quality of output were improved. The greatest increases in productivity were reported by users in departments outside human resources where individuals with relatively little tenure in the organization are direct computer users. Human resource programs which will maximize the positive impacts of information system utilization include job redesign, modified reward systems, individual and group training, and appropriate communication systems. Managers who are sensitive to the impacts of new computer technology and develop such programs may maximize the effective use of information systems.

Introduction

In 1982 the Human Resource Planning Society Roundtable identified a number of key issues for the 1980s. These included the impact new computer technology will have on productivity in general and on job design, reward systems, and other management areas in particular (Critical Human Resource Issues of the 1980's, 1983). Past research has addressed some of these issues, and the current study extends this work to a sample of users of human resource information system (HRIS) technology. The HRIS used in many organizations is similar in design, content, and applications, thus allowing inter-organizational compari-

* The research on which this article is based was funded in part by the Midwest Human Resource Planners Group and the Human Resource Planning Society.

sons to be made. In addition, the HRIS provides information to managers throughout the company, potentially impacting all areas of the firm.

Recent surveys show that the majority of large companies have some type of HRIS available for basic payroll, record-keeping, and administrative tasks (*Magnus* and *Grossman* 1985, *Mathys* and *LaVan 1982, Mathys, LaVan,* and *Nogal* 1984). In addition, specialized tasks such as compensation, benefits, equal employment opportunities (EEO), applicant tracking, training administration, labor relations, and human resource forecasting have been automated (*Amico* 1981, *Hoff* 1983, *Lapointe* 1983, *Weirmar* 1980).

The mode of access to these systems has shifted in recent years. Retrieval of information was often centralized and required a relatively lengthy process of requesting a special report, waiting for it to be "programmed" by a specialist, and receiving a final report. Users today often expect to have "hands on" access to the human resource information system, with retrievals done by themselves or their subordinates. This reflects the general shift to widespread use of microcomputers and computer terminals by "end users," those who actually utilize the information obtained *(Giuliano* 1982). Within human resource departments, a phone survey completed in late 1984 indicated that in 89 percent of 81 midwestern companies included, human resource professionals and managers were using the computer themselves (*Verdin* and *Pagano* 1987). In other areas, such as accounting and finance, the utilization rate has been found to be even higher than in human resources (*Quillard* et al. 1983).

The present study examines the attitudes of information system users from human resource departments regarding the impact of computer technology on their jobs and careers. Also, the attitudes of human resource department users versus users from other departments about the impact of information system utilization on decision making, productivity, and quality of output are compared. In addition, the relation of organization and end-user background characteristics to quality of output and productivity are examined.

Literature Review

A number of studies have examined the changes in skill requirements of jobs after the introduction of computer technology. Reduced skill requirements have been reported by both clerical workers and managers (*Harris* and *Harris* 1982, *Zuboff* 1982). However, an increase in skill requirements has been reported by office automation users in 55 offices in 26 organizations (*Gutek, Bikson*, and *Mankin*, 1984). Interviews with personal computer users indicated that although their work has changed, improved performance and quality of the work has resulted (*Quillard* et al. 1983).

As computer skills are acquired and employees are able to use the new technology to improve their performance, a positive impact on career opportunities may result. Not only are personal computer skills becoming a requirement for many professional jobs, but new careers such as manager of decision-support systems or micro-computer applications are being developed. In addition, one study indicates that there is a positive relationship between individual perceptions of career success and attitudes toward the quality of job life and work effectiveness when computers are used *(Gattiker* and *Coe* 1986). Those who see the computer as affecting their jobs in a positive manner also see their careers as successful.

One of the primary purposes of an information system is to provide managers with information which supports the decision making process. In order to assess the perceived impact of computer technology on decisions, three questions were developed to measure changes in the time spent on making decisions, the amount of data used, and the quality of the decisions made (see *Allen* and *Emmelhainz* 1984). Although this area has received theoretical treatment, empirical studies about the impact of information technology on the decision making process have been lacking. These items were answered by all HRIS users.

In addition, it has been found that, in general, the utilization of computer technology has been perceived to positively impact job performance. Changes include: increased speed of the work, improved quality of performance, and new analyses leading to better understanding of the work (*Gutek* et al. 1984, *Quillard* et al. 1983, *Verdin* 1987, *Verdin* and *Pagano* 1987).

Finally, individual differences in background and experience with automated systems may be related to the extent to which new technology is used and an individual's disposition toward it (*Gattiker* 1984, *Gattiker* and *Nelligan* 1988, *Staw*, *Bell*, and *Clausen* 1986). Organizations also have varying policies regarding information systems. These differences may be reflected in the design of the information systems, the amount and type of equipment available to end-users, the reward structures associated with effective utilization of the technology, and the ultimate success of the implementation (*Pagano* and *Verdin* 1987).

Research Hypotheses

The research reported here addresses several specific hypotheses regarding the perceived changes resulting from the availability of human resource information systems and computer technology to end-users. These include:

1. The utilization of an information system will increase the skill requirements and inter-departmental communications of users in the human resource department.

Past research indicates that jobs generally change in the direction of increased skill requirements, especially for managers and professionals who may not have been using computer technology directly (*Gutek, Bikson,* and *Mankin*, 1984). In addition, interaction with computer support personnel and other information system user departments makes increased communication likely.

2. The utilization of an information system will increase the perceived career opportunities of human resource department users.

As increased skills are acquired and the applications for which computers are utilized expand, individuals may perceive increased career opportunities (*Gattiker* and *Coe* 1986). Also, improved productivity and quality of work may indirectly lead to increased career opportunities.

3. The utilization of an information system by all HRIS users will positively affect the decision making process.

Increased access to information could result in faster, better quality decisions. Thus, the availability of human resource information systems to managers and professionals is hypothesized to result in the perception of improved decision making processes (*Allen* and *Emmelhainz* 1984).

4. Utilization of an information system will increase the perceived productivity and quality of output by all HRIS users.

The increase in productivity and quality of work as a result of using computer technology has consistently been shown to be the case in previous research (*Quillard* et al. 1983, *Verdin* 1987, *Verdin* and *Pagano* 1987). The various applications of computer technology, from office automation to personal computer applications, provide users with a resource which allows them to do their jobs better.

5. Users outside the HR department will report more changes in the decision making process and greater increases in productivity and quality of output than human resource department users.

Computer information systems have been used in areas such as finance, marketing, and accounting for many years (*Quillard* et al. 1983). Because the users in these areas have had greater experience, and more of their information resources may be automated, than individuals in human resource departments (e.g., industry financial data), it is probable that improved decision making and greater productivity and quality of work improvements will be perceived after the information system is introduced.

6. The quality of output and productivity correlate significantly with organizational context and end-user background variables including supervisory responsibility, tenure with the company, type of department, and computer experience.

An individual's background has been found to impact his or her perceptions of the technology being used (*Gattiker* 1984, *Gattiker* and *Nelligan*, 1988, *Staw, Bell,* and *Clausen* 1986). In addition, the success of the implementation of new technology as reflected in improved quality of output and increased productivity may also be related to the organizational policies and context within which the information system exists (*Pagano* and *Verdin* 1987). Thus, a significant relationship between individual and organizational factors and perceived quality of output and productivity is predicted.

Method

Design

The research questions outlined above were tested using a sample of human resource managers and professionals and other managers who are using recently updated or installed human resource information systems. (Managers are defined as individuals who have some direct supervisory responsibility. Professionals or specialists do not directly supervise others, but hold jobs above the clerical or support staff level.) Because of the similarity of the data collected and the applications implemented, the impact on HRIS users can be compared across companies. In addition, computer technology has been recently adopted by many human resource managers. Therefore, attitudes about changes in job skills, career expectations, decision making, productivity, and quality of output can be more readily ascertained than in areas where such technology has been used for many years.

The data collection was carried out using a two step process. First, key human resource information system personnel from three organizations were interviewed with regard to the characteristics of the information system used by their departments and the availability of hardware and software to users. In this way it was possible to ascertain the level of usage, the general types of applications, and to get approval to administer a follow-up questionnaire. Second, a questionnaire was given to HRIS users both from within the HR department and from other departments.

Individuals from three organizations were included in the sample. In two cases, all human resource personnel were included. In the third organization, only the human resource department users who had access to the recently installed system (approximately 50 percent of the members of the department) received questionnaires. In addition, the managers outside the human resource department who request or receive information from the system were identified and sent questionnaires through the company mail. In all cases, questionnaires were returned directly to the researchers, and all responses were kept anonymous.

Organizational Characteristics

The organizations in the sample were from the financial, manufacturing, and transportation industries. Interviews with individuals directly responsible for the human resource department information system indicated that basic administrative functions were automated as were several specific areas such as compensation, administration of training programs, and applicant tracking. In one organization, the basic information system had been available for many years, but direct user access was relatively new. All of the organizations had microcomputers and/or computer terminals available to the users. It appeared that those individuals who could justify the use of the computer for their current or future work were able to acquire the necessary equipment and programs.

Characteristics of the Sample

A total of 107 usable questionnaires were returned during Fall, 1985. This represents a 53 percent response rate. Of these, 74 (69 percent) were employed in the HR department while the remaining 31 percent were from other areas. Characteristics of HRIS users are shown in Table 1. Positions held ranged from support personnel to Vice President or Director. Most of the respondents from the human resource department were supervisors/specialists, administrators, or managers. The majority of users from other departments were supervisor/specialists, managers, or Vice President/Directors. Individuals outside the HR department had more supervisory responsibility than human resource department users. The average human resource department user had been with the organization for 9.2 years and in their current position for 3.9 years. The other users reported an average time in the organization of 13.9 years and had spent 6.1 years in their present position.

The majority of both groups reported using the computer themselves at least part of the time. However, two-thirds of the users outside of the human resource department had been using the computer for more than two years, while three quarters of the human resource department users had less than two years experience with this technology. Most of the users in both groups reported spending less than 10 percent of their time at terminals or personal computers. The technology was apparently being used as an additional tool for completing their regular work.

The differences between these two groups reflect the fact that computer technology has been used by individuals in departments such as finance, marketing, and accounting longer than in human resource departments. This is similar to the results reported by *Quillard* et al. (1983).

Table 1: Background Characteristics of HR Department Users Versus Users from Other Departments

Characteristics	*HR Users*		*Users Outside HR*	
Position	N	(%)	N	(%)
Support personnel	12	(17.1)	1	(3.0)
Supervisor/specialist	27	(38.6)	8	(24.2)
Administrator	10	(14.3)	1	(3.0)
Manager	17	(24.3)	10	(30.3)
Asst. VP/Division Manager	3	(4.3)	1	(3.0)
VP/Director	1	(1.4)	12	(36.4)
Number of Employees Supervised				
None	40	(55.6)	4	(12.1)
1–10	10	(38.8)	15	(45.3)
More than 10	4	(5.6)	14	(42.6)
Computer User				
Yes	47	(63.5)	25	(75.8)
No	27	(36.5)	8	(24.2)
For Computer Users, Time Computers Have Been Used				
Less than 1 year	7	(15.6)	1	(4.0)
1–2 years	27	(60.0)	8	(32.0)
More than 2 years	11	(14.4)	16	(64.0)
For Computer Users, Percent of Time Computers are Used				
10% or less	30	(66.7)	20	(80.0)
More than 10%	15	(33.3)	5	(20.0)

Measures

Items relating to decision characteristics, the impact of computer technology and human resource information on performance, and the change in the quality of their work were included in the questionnaires to all respondents. In addition, individuals in the human resources department (69 percent of the sample) were asked how the computer impacted their jobs, skills required, interdepartmental communications, and their careers. The items were scaled from 1=low to 5=high. These items were developed from the literature and from the results of interviews done in a previous study (*Verdin* and *Pagano* 1987). The specific items used along with their means and standard deviations are presented in Table 2.

These measures attempt to tap the attitudes and perceptions of information system users. Single items were used to increase the return rate and to allow more areas to be covered in the questionnaire. An attempt was made to incorporate several dimensions of performance and decision making characteristics into the questionnaire.

Table 2: Item Means and Standard Deviations (1 = low; 5 = high)

Item	*N*	*Mean*	*S. D.*
1. Extent of job change in general	74*	3.2	.9
2. Extent job has become de-skilled	72	1.5	.9
3. Extent skill requirements have increased	73	2.9	1.1
4. Communication with other departments	70	3.7	.7
5. Enhanced career opportunities in firm	73	3.3	.6
6. Enhanced career opportunities outside firm	72	3.5	.7
7. Time spent on decision making	106	3.2	.5
8. Amount of data used for decision making	104	3.5	.8
9. Quality of decisions	105	3.5	.6
10. Productivity since having information	105	3.7	.7
11. Productivity since computer technology	105	4.0	.7
12. Quality since computer technology	105	4.0	.7

* Only the human resource department users answered questions 1–6.

Results

Changes in Job Skills and Communications

Hypothesis 1 stated that utilization of an information system will increase the skill requirements and interdepartmental communications of human resource managers and professionals. Eighty-five percent of the human resource department employees in this sample indicated some changes in their jobs had occurred since computer technology had been introduced. Forty-eight percent indicated the use of the computer to monitor performance of employees had increased, and 57 percent indicated communication and interaction with non-HR employees had risen. In general, the respondents do not do work at home on personal computers or terminals.

With respect to skill requirements, only 19 percent of the respondents indicated their jobs had been deskilled (i.e., the skill levels required to do the job had been reduced). However, 61 percent felt their job requirements had been increased. Thus, for the human resource department users in this sample, job changes appear to be in the direction of increased skills, computer monitoring, and interaction with other departments. The first hypothesis is supported for respondents in human resource departments.

Changes in Career Opportunities

It was hypothesized (H2) that the utilization of an information system will increase the perceived career opportunities of human resource department users. In the present study, individuals working in human resource departments were asked how the introduction of the computer has changed their career opportunities both inside and outside the organization. Thirty-five percent of the sample indicated that gaining the skills needed to use this technology enhanced their career opportunities within their organization. However, 49 percent felt their career opportunities would be enhanced in other organizations. The majority of the remaining respondents felt there would be little or no impact on their careers.

If companies do not establish appropriate rewards and opportunities for the employees trained to use computer technology, some may choose to further their careers by leaving the organization where the skills were acquired to join companies which do reward this knowledge. The perception of increased career opportunities outside the organization was greater than improved promotional opportunity within the firm. The hypothesis was partially supported for this sample.

Improved Decision Making Processes

Hypothesis 3 stated that increased utilization of a human resource information system by managers and professionals in a firm will positively affect the decision making process. Twenty-three percent of the human resource department users and 21 percent of the users from other areas reported spending less time on decision making, leaving more time for other tasks. In addition, 42 percent of the human resource department users and 36 percent of other users reported using an increased amount of information for decision making. Less than 4 percent of the respondents reported using more time or less data for these activities. It appears that when changes in the decision making process were reported, more information was utilized, and time was spent more efficiently.

With respect to quality of decisions, 40 percent of the human resource users and 42 percent of the other users indicated increases in quality. Decision quality remained the same, according to the remaining respondents. There appeared to be some support for this hypothesis, although the results are mixed.

Improved Productivity and Quality of Output

An increase in perceived productivity and quality of output was predicted for all HRIS users (H4). Seventy-one percent of the human resource users and 91 percent of the users from other departments reported increases in their productivity as a result of computer technology. In addition, 64 percent of the human resource department users and 42 percent of the other respondents indicated increased productivity as a result of having human resource data available. For the majority of respondents, increases in productivity were found. Computer technology in general had increased productivity for almost all users from other departments, although the availability of HR information had a less dramatic impact. These results support previous research.

Just as computer technology can improve productivity, it has been found that the perceived quality of output or work performed also improves (*Gutek* et al., 1984, *Quillard* et al. 1983, *Verdin* and *Pagano* 1987). Sixty-eight percent of the human resource department users and 88 percent of the users from other areas felt the quality of their work improved. Hypothesis 4 was supported for this sample.

Human Resource Versus Non-HR Managers

Hypothesis 5 indicated that users from other departments will report significantly greater improvements in decision making processes, productivity, and quality of output than human resource department users. One-way analysis of variance was used to determine if significant differences existed with respect to the decision making process, productivity, and quality of output for these two groups. The results are presented in Table 3.

There were no significant differences between users from human resource departments versus other areas regarding the time spent on decision making, the amount of data used, and the quality of decisions. However, significant differences were found with respect to productivity and quality of output. Users outside the human resource department reported significantly higher productivity as a result of computer technology in general and increased quality of output. On the other hand, human resource department users reported significantly higher productivity as a result of having human resource data available. It appears that gains as a result of changes in technology are more prevalent in departments outside of human resources except where the information is specific to this area.

Table 3: One-Way Analysis of Variance of Human Resource Department Users Versus Users from Other Areas

Item	*N*	*Mean*	*S. D.*	*F*
Time spent on decision making				
Human Resource Users	73	3.2	.5	N. S.
Other Users	33	3.2	.5	
Amount of data for decisions				
Human Resource Users	71	3.5	.9	N. S.
Other Users	33	3.4	.8	
Quality of decisions				
Human Resource Users	72	3.5	.6	N. S.
Other Users	33	3.5	.6	
Productivity since having HRIS				
Human Resource Users	72	3.8	.7	4.5*
Other Users	33	3.5	.7	
Productivity since computer technology				
Human Resource Users	72	3.8	.6	22.9***
Other Users	33	4.5	.7	
Quality of output since computers				
Human Resource Users	72	3.8	.7	21.7***
Other Users	33	4.5	.7	

* $p < .05$
*** $p < .001$

Impact of Organization and Individual Background

Further analysis was done to test hypothesis 6, which predicted that organization and end-user background characteristics would explain differences in the impact of information technology on productivity and quality of work. These characteristics included position in the organization, the number of employees supervised, computer user or nonuser, number of years in the organization, and number of years in their present position.

Models were successfully developed predicting productivity and quality of output as a result of computer technology in general. These individual characteristics did not predict decision quality or productivity as a result of having human resource information available.

Table 4 describes the model predicting individual productivity as a result of having computer technology available. The model explains 40 percent of the variation in the dependent variable. The overall regression is significant at the

.0001 level. Variables predicting relatively high productivity are membership in Organization A, supervisory responsibility, and being a computer user. In addition, users outside the human resource department and those with the organization a relatively short period of time indicated greater productivity.

Table 4: Regression Model Predicting Productivity Since Computer Technology
$R^2 = .40$
$F = 12.79$ ($p < .0001$)
Productivity Since Computer Technology Introduced

	β *Value*	*t*	*p*
Intercept	4.06		
Organization A	.34	2.71	.008
Other Users/HR Dept Users	–.58	–4.29	.0001
Number subordinates	.01	2.76	.006
User/Non-User	.50	3.99	.0001
Time in organization	–.02	–3.10	.002

Table 5 describes the model predicting increases in quality of output. The model explains 34 percent of the variation in the dependent variable. Again membership in Organization A predicts greater levels of quality of output. Also, computer users, those with supervisory responsibility, users outside the human resource department, and those who had been with the company a relatively short time indicated higher quality increases.

Table 5: Regression Model Predicting Quality of Output Since Computer Technology
$R^2 = .34$
$F = 9.82$ ($p < .0001$)
Quality of Output Since Computer Technology Introduced

	β *Value*	*t*	*p*
Intercept	4.10		
Organization A	.29	2.12	.036
Non-HR/HR	–.57	–3.87	.0002
Number subordinates	.01	2.96	.003
User/Non-User	.39	2.85	.005
Time in organization	–.02	–2.52	.013

Both models reflect the importance of the system itself in predicting productivity and quality of output. Members of Organization A had the human resource information system available for many years and had only recently gained direct user access to it. The system was well established and users were familiar with what information was available. Direct access gave them decreased turnaround time and possibly more flexibility in how the information could be retrieved. This apparently was reflected in higher productivity and quality of output.

As indicated by the regression analysis, users outside the human resource department felt their productivity and quality of output was improved more than human resource department users and professionals (Beta Value was -.6 in both cases). In addition, those individuals who were relatively new to the organization, who supervised a number of individuals, and who used the computer directly, expressed greater increases in these areas.

Discussion

The purpose of this study was to examine the impact of information technology on the perceived changes in jobs and career opportunities for a sample of human resource department users. In addition, the impact of this technology on decision making, productivity, and quality of output was studied for a more general sample of end-users of the human resource information systems in three organizations. Finally, the relationship between organization and individual background and productivity and quality of output was studied.

In general, previous research was supported. As reported by *Gutek, Bikson,* and *Mankin* (1984) and *Quillard* et al. (1983), jobs changed in the direction of increased skill requirements, increased computer monitoring, and increased interaction with other departments. Further research is needed to study specific jobs before and after information technology is implemented to objectively determine the extent of such changes.

With respect to career opportunities, the organizations studied did not appear to have a promotion or reward policy dealing directly with the skills acquired by computer users. More respondents perceived improved career opportunities outside their current organization than within it. Again, further research is needed to trace actual career paths of individuals who have effectively used information technology. It is possible that, in some contexts, individuals will reduce their career opportunities if they "spend too much time at the computer" rather than doing their job.

Although a great deal of discussion centers around the use of information systems to support decision making, relatively little has been done to examine just what this impact is. The results of this study indicate that the increased amount of information is being used by many managers, the quality of decisions is perceived to be better, and (for about one-fifth of the sample) less time is devoted to decision making. Although the results are encouraging, much work needs to be done to try to determine if the quality of the decisions has actually improved after information systems are available, which managers can use these systems most effectively, and ways the information contained in the systems can be accessed to best advantage (*Verdin* and *Lapointe*, in press). This is perhaps one of the most challenging areas for future research.

The results of the study also support past research which reported improved productivity and quality of output (*Gutek, Bikson*, and *Mankin* 1984, *Quillard* et al. 1983, *Verdin* 1987, *Verdin* and *Pagano* 1987). Increases were reported in these areas by end-users in all departments sampled. However, some individuals reported little or no change in productivity and quality of output. The reasons for this are many (see *Pagano* and *Verdin* 1987) and should be examined in a variety of situations by future researchers.

Finally, individual differences in background and organization were examined with respect to the managers' perceptions of productivity and quality of output. It was found that the users of the system in Organization A reported higher productivity and quality of output than users from Organizations B and C. Although it seems likely that the organization variable is a dummy or surrogate for length of time the system is in place, future research will be needed to establish this with any certainty. Individual characteristics which were significantly related to productivity and quality of output were short tenure in the company, supervisory responsibility, direct computer users, and membership in departments outside of human resources. Future research which examines similar relationships for many users of one system would be able to focus on individual characteristics, controlling for the system being used. Other import characteristics such as gender (see *Gattiker* and *Coe* 1986, *Gattiker* and *Nelligan* 1988) should also be examined.

Conclusions

Computer technology does appear to have several direct impacts on users of human resource information systems. Job changes did occur, especially in the direction of increased skill requirements and a greater amount of communication with members of non-HR departments. De-skilling was not a particular problem for these respondents.

Second, career opportunities are not necessarily linked to learning or using new information technology. This is particularly true within the organization, although a greater percentage of individuals felt that this knowledge would be useful in obtaining a job with another firm. However, since the respondents felt productivity and quality of work was improved, those individuals who were successful information system users may be more likely to be rewarded over the long term.

Third, there is a trend toward increased quality in decisions as reported by the respondents; however, there is no difference between human resource department users and users from other areas regarding this variable. Further research into the decision making process before and after a computerized information system is installed is needed. The process may remain the same, but the increased availability of information and the enhanced analysis capabilities of automated systems may affect the quality of the decisions made.

Finally, both productivity and quality of output were reported by the respondents to have increased. In addition to differences between human resource department users and information system users from other departments, predictors of increased productivity and quality of output were affiliation with Organization A, number of employees supervised, direct computer usage, and a short time in the company.

Managerial Implications

Changing job and communication requirements may result in the need for job redesign and communication flows. In order to insure that implementation problems are kept to a minimum and productivity increases maximized, managers can examine the workflow and job design before and after a new information system is installed or modified to make any changes necessary (*Pagano* and *Verdin* 1987). New job specifications, job grades, and compensation schedules may be required to reflect the situation after the new information technology has been implemented.

Appropriate training programs should also be made available to users and potential users to make sure the computer system is being used properly and to enhance productivity and quality of output to the greatest extent possible. A combination of classroom training, individual instruction, and time and equipment made available for programmed instruction might be appropriate. Since communication across departments increases after computer technology is introduced, managers might want to take advantage of the potential cross-fertilization of ideas. User groups, suggestion boxes, newsletters, and other formal means of communication may make employees in various departments aware of potential applications, mutual problems, and available software.

Companies investing in the hardware and software which allows users to retrieve and analyze information from computerized data bases should consider the rewards available to those who obtain the necessary increased skills to effectively utilize the new technology. If employees are not rewarded, they will be likely to minimize their use of the technology, and potential productivity gains will be lost. Incorporating the use of information technology into existing compensation and promotion systems will increase the likelihood that the new technology will be used appropriately.

Managers can expect the greatest productivity and quality increases in departments outside of human resources where users with relatively little tenure in the organization are direct computer users. As these individuals move up through the organization, it is possible that their utilization of computer technology will be emulated and the benefits of information systems will spread.

References

Allen, M. K. and *Emmelhainz, M. A.* (1984): Decision support systems: An innovative aid to managers, *Journal of Business Logistics*, 5 (2): 128–142.

Amico, A. M. (1981): Computerized career information, *Personnel Journal*, 60 (August): 632–633.

Critical human resource issues of the 1980's (1983): *Human Resource Planning*, 6 (2): 63–67.

Gattiker, U. E. (1984): Managing computer-based office information technology: A process model for management, in: *H. W. Hendrick* and *O. Brown*, Jr. (Eds.), *Human factors in organizational design*, Amsterdam: Elsevier Science, 395–403.

Gattiker, U. E. and *Coe, L.* (1986): Relationship of computer attitudes with perceptions of career success, In: *J. A. Pearce* II and *R. B. Robinson* (Eds.), *Best paper proceedings of the 46th Annual Meeting of the Academy of Management,* 37: 294–298.

Gattiker, U. E. and *Nelligan, T. W.* (1988). Computerized offices in Canada and the United States: Investigating dispositional similarities and differences, *Journal of Organizational Behavior*, 9: 77–96.

Giuliano, V. (1982): The mechanization of work, *Scientific American* (September): 149–164.

Gutek, B. A., Bikson, T. K., and *Mankin, D.* (1984): Individual and organizational consequences of computer-based office information technology, in: *S. Oskamp* (Ed.), *Applied Social Psychology Annual,* Beverly Hills, CA: Sage Publications, 5: 231–254.

Harris, P. R. and *Harris, D. L.* (1982): Human resources management, part 1: Charting a new course in a new organization, a new society, *Personnel*, 59 (5): 11–17.

Hoff, R. D. (1983): The impact of cafeteria benefits on the human resource information system, *Personnel Journal*, 62 (April): 282–283.

Lapointe, J. R. (1983): Human resource performance indexes, *Personnel Journal*, 62 (July): 545–553.

Magnus, M. and *Grossman, M.* (1985): Computers and the personnel department, *Personnel Journal*, 64 (April): 42–48.

Mathys, N. and *LaVan, H.* (1982): A survey of the human resource information systems (HRIS) of major companies, *Human Resource Planning*, 5 (June): 83–90.

Mathys, N., LaVan, H., and *Nogal, G.* (1984): Issues in purchasing and implementing HRIS software, *Personnel Administrator*, 29 (August): 91–97.

Pagano, A. M. and *Verdin, J. A.* (1987): Implementing computer technology in transportation, *Transportation Quarterly,* 41: 381–396.

Quillard, J. A., Rockart, J. F., Wilde, E., Vernon, M., and *Mock, G.* (1983): *A study of the corporate use of personal computers*, Center for Information Systems Research, Working Paper 109. Unpublished manuscript, Massachusetts Institute of Technology.

Staw, B. M., Bell, N. E., and *Clausen, J. A.* (1986): The dispositional approach to job attitudes: A lifetime longitudinal test, *Administrative Science Quarterly* 31: 56–77.

Verdin, J. A. (1987): Up and down the HRIS career ladder, *Computers in Personnel*, 1 (Spring): 35–39.

Verdin, J. A. and *Lapointe, J. R.* (in press): Performance measurement and decision evaluation cost models using human resource data, *Personnel Journal.*

Verdin, J. A. and *Pagano, A. M.* (1987): Current trends in the use of computer technology by human resource managers, In: *R. J. Niehaus* (Ed.), *Strategic human resource planning applications*, New York: Plenum Press.

Weirmar, K. (1980): Objectives and measures of coordination between private and public manpower planning with particular reference to the use of human resource measurements systems, *International Journal of Manpower (UK)*, 1: 11–16.

Zuboff, S. (1982): New worlds of computer-mediated work, *Harvard Business Review*, 60 (5): 142-152.

Chapter 8
Computer End-Users: The Impact of Their Beliefs on Subjective Career Success*

Urs E. Gattiker

Abstract

This paper examines a person's positive or negative beliefs about computerization of the office and his/her perception of career success. The results suggest that such beliefs about computers relate to subjective evaluations of job success and general career success. Respondents varied according to category of computer used most often (intelligent workstation versus main-frame terminal). The data also showed that the perceived effect of the computer's control over the individual worker differs strongly according to gender. The implications of these results for practitioners are discussed, and guidelines for researchers who develop theories about career success and computer exposure are provided as well.

The introduction of computer-based technology into offices is changing the workplace of North America. This change is fuelled by societal and organizational requirements for increased effectiveness and profitability. Policy and strategy issues as related to computer-based technology have been central in recent organizational literature (*Megaw* and *Lloyd* 1984), while human resource aspects have received very little attention from organizational researchers (*Kahn* 1981). Yet, in a society where career success is of paramount importance, the individual's assessment of, and beliefs about, technological change in office settings are crucial, for they may well affect his/her career prospects (*Rosenbaum* 1985).

This paper investigates how an individual's perception[1] of his/her career success might be related to his/her beliefs[2] about a computer-based technology[3]. Specifically, this study looks at the construct of career success and its facets,

* Financial support for this research project was provided by the Social Sciences and Humanities Research Council of Canada and The University of of Lethbridge Research Fund, Contract No. 86–1934–405, by a grant from the Alberta Manpower "STEP" Program, Contract Nr. 11149, and by the Alberta Manpower Post-Secondary Training Program. Many thanks to Larry Coe and Todd Nelligan for their help with data collection and analysis.
Correspondence concerning this article should be addressed to Urs E. Gattiker, School of Management, The University of Lethbridge, Lethbridge, Alberta, T1K 3M4, CANADA.

such as job and financial success, as perceived by the individual. It will also investigate how these perceptions are coupled with one's beliefs about computers and an assessment of worker's views about quality of job life and communication. Any differences in employee beliefs about computer-based technology due to the type of equipment used, hierarchical level in the organization and gender will be determined as well.

Literature Review

The approach presented here is inter-disciplinary, drawing upon the literature of sociology,. management, and psychology. Studies in three specific areas have a direct bearing on this research: work attributes, career development as well as conceptualizations of computerization and its effects upon the workplace. A brief summary of recent work in these areas will be presented to show how it can be applied to the office technology domain.

An Individual's Perception of Career Success

Individuals attempt to be successful in their careers (*Bandura* 1982), but some data indicate that so-called "successful" managers may not perceive themselves as being such. Objective measures of success have generally failed to explain attitudinal outcomes, and thus they may not explain the construct of career success very well (*Korman* 1980). People's success constructs are influenced by their environment, which includes peers, friends and family (*Romney* et al. 1979). As a result, an individual's own construct is not totally independent of society's definition of success.

One recent research project (*Gattiker* and *Larwood* 1986) developed a scale for measuring perception of career success of white-collar workers in metropolitan areas in the U. S. Four facets of perception of career success were measured: financial, interpersonal, job, and general career success. The data revealed that job and general career success were especially important for support staff, which seems rather surprising in a society which puts great importance on material objects and interpersonal relationships.

The connection between perception of career success and individual beliefs about technology is unclear. Research by *Gattiker* (1987), using an expanded version of the above measures, showed that a person's perception of financial and interpersonal success has nothing to do with his/her attitude towards a specific technology, while feelings of job success and general career success indeed related to the respondent's beliefs about computer-based technology.

Individual Beliefs About Working with a Computer

Fishbein (1967) suggested that one useful way to conceptualize the notion of attitude is to subdivide it into three related parts, namely (1) a person's beliefs about job and work, (2) the attitude itself (e.g., dissatisfaction), and (3) behavioral intentions. The focus in this paper is on the first part, testing if one's beliefs about computer-mediated work and perception of career success are related to each other. Unfortunately, researchers have seldom made an attempt to assess individual belief states of employees who have several years of computer experience (*Gutek* 1983).

The individual's cognitive evaluation of computers in the workplace could help explain differences in beliefs. A positive belief about working with computers is desirable, since some researchers have argued that maximum effectiveness of any technology requires users to feel comfortable with it. For example, one recent study showed that personal computers were perceived as being most helpful in improving work effectiveness and the quality of job life (*Gattiker, Gutek*, and *Berger*, in press). Other relevant concerns are the technology's impact upon one's performance as well as its perceived control over employees (*Salzman* and *Mirvis* 1985). Although computers may be given credit for improving efficiency and effectiveness on the job, they can also provide new avenues of control for management, which could lead to worker hostility and unrest (*Mankin, Bikson,* and *Gutek* 1982). Very little empirical research, however, has been done in this area (e.g., *Kling* and *Iacono* 1984).

Still another related factor is the individual's belief about the computer's effect on communications inside and beyond the firm. Some technologies are more useful for this purpose than others, but the computer appears to have become an important new tool for improving communications. Yet, research has shown that people communicating via computers evaluated each other less favorably than did people dealing face-to-face (*Kiesler* et al. 1985). This result is of crucial significance, since a major part of office work involves the exchange of large amounts of data and information (*Doswell* 1983, *Panko* 1984).

Responses According to Computer Technology and Demographics

Computer technology. Some researchers have argued that different types of computer-based technology affect people's work differently (e.g., *Salzman* and *Mirvis* 1985). When compared to users of personal computers or word processors, employees working with main-frame computers may perceive greater control being exercised by the technology (cf. *Kling* and *Iacono* 1984). There is also the fear that their work is being de-skilled and that their job within the organization may become obsolete.

Much attention has been focused on intelligent workstations and their effectiveness in facilitating communication both within and beyond the organization (*Demby* 1985). Spreadsheet and word-processing programs allow preparation of highly presentable reports, often including graphics and figures. These and other new capabilities can improve communication through more effective use of the medium from either side (*Pava* 1983). Again, applied organizational research studying computers and communication is limited, and most existing research has not compared responses from individuals working with main-frame terminals as opposed to intelligent workstations (*Kiesler* et al. 1985).

Gender. Various research data showed that women are often more affected by computerization than men due to their occupations and positions, but these studies have not examined specifically if women evaluate the technology itself differently than men do (*Form* and *McMillen* 1982, *Gutek* 1983). Therefore, a need exists for more research investigating how gender affects differences in beliefs (*Kling* 1978).

A simple examination of sex differences, however, might be misleading. Men and women may use the same technology, but in different positions and thus for different purposes. For example, in traditionally female clerical jobs, the computer may dominate and define the job (e.g., secretary or data-entry operator), while in professional and managerial positions, the computer may be one of many useful tools which assist in the completion of varied tasks (*Gutek* and *Larwood*, in press).

A study by *Gattiker, Gutek* and *Berger* (in press) looked at gender as a factor across similar positions. Respondents were selected from various levels and, most importantly, women were selected first and men second. This approach was used to ensure that both men and women would have similar duties and tasks to perform. The data revealed that women perceive computers less positively than men do. These differences were even more apparent when comparing respondents who work with personal computers. Research by *Gattiker* and *Nelligan* (in press) found that women were very concerned about the effect of the technology upon their quality of job life. While men and women *both* felt main-frame terminals affected their communication patterns negatively, men thought personal computers improved communications, while women were less likely to feel that way.

It appears that even though changes in the workplace have occurred (*Form* and *McMillen* 1982), they are not necessarily perceived as such by employees. Furthermore, a comparison of women and men in *similar positions* reveals that differences in perception do exist between the sexes. *Morgall's* (1983) personal account of experiencing computerization in an office setting is helpful, but work investigating technology's impact on beliefs needs to go further. Predicting future effects of technological breakthroughs on women and their work is important (*Gutek* and *Larwood*, in press), but the identification of past and current effects should be of even greater significance (*Gattiker*, in press).

Hierarchical level. While technologies may change work content at various levels, beliefs would not necessarily reflect such developments (*Gattiker* 1984). Individual perceptions of computers in the workplace might not necessarily differ along hierarchical lines. This does not mean that computerization affects managers and support personnel similarly. Instead, a self-selection process may have eliminated objective differences in job content (*Gattiker* and *Larwood* 1986, *Spenner* 1983). One explanation for the lack of differences according to organizational levels could be that people tend to seek jobs which offer a structure compatible with their aptitudes and skills (*Kohn* et al. 1983, *Spenner* 1983).

One recent American study showed that managers did not differ from support personnel in how they perceived their computers (*Gattiker, Gutek,* and *Berger*, in press). Similar results are reported by *Gattiker* and *Nelligan* (1988) about a Canadian sample. These studies indicate that although workplace effects may be very different for managers and support personnel, their beliefs about the technology do not necessarily differ. This issue should be examined more closely, since lack of differences in people's beliefs would further substantiate the argument that people select jobs which fit their needs (*Brousseau* 1983).

Summary

A variety of factors influence individuals' beliefs about working with computers and their perception of career success. Demographic factors, such as gender, hierarchical level, and the type of computer used (main-frame versus intelligent workstation), could help distinguish end-users and their beliefs about computers. Researchers have pointed out that perception of career success is an important factor in most developed countries in assuring a person's well-being (cf. *Bandura* 1982, *Kahn* 1981). Technological innovation in the workplace, however, could affect subjective career success. Furthermore, efficient use of computers requires that employees approach them with a positive attitude (e.g., *Carter* 1984).

Research Issues

This study examined perception of career success with the following four predictors: an individual's beliefs about computer-based technology as it pertains to quality of job life, work effectiveness, communication, and control. The expectations to be tested here are that positive correlations exist between a person's beliefs about computers and one's perception of job and general career success. People have an inherent desire to be successful in what they are doing (e.g., *Bandura* 1982, *Gattiker*, in press, *Korman* 1980), while more and more people are or will be affected by computerization.

Can the type of computer used result in varying employee beliefs about the technology, and can gender and hierarchical level in the organization affect those beliefs? Various researchers have mentioned these factors as potentially important, so several analyses will be conducted to investigate these issues (e.g., *Form* and *McMillen* 1982, *Salzman* and *Mirvis* 1985). Additionally, a scale developed previously to measure perception of career success (*Gattiker* and *Larwood* 1986) and an expanded version of a measure to assess computer beliefs (*Gattiker, Gutek*, and *Berger*, in press) will be used to evaluate their applicability more widely.

The following hypotheses were established:

Hypothesis 1. In agreement with the literature previously cited, quality of job life, work effectiveness, communication, and control will explain a significant part of the population variance and correlate positively with one's perception of *job success* and *general career success.*

It was specifically predicted that:

H1a. Perceived quality of job life as well as work effectiveness will explain the largest part of the population variance in perception of job success and general career success of any predictor set used in this study.

H1b. All predictor variables together will explain a significant part of the population variance in job success and general career success.

Judging from previous organizational research, the usefulness of beliefs about computers for explaining a person's perceived financial and interpersonal career success was open to question (cf. *Gattiker* 1984, *Spenner* 1983)..

Hypothesis 2. In accordance with the literature cited earlier, computer effects on communication and control will be perceived differently by users of non-intelligent versus intelligent workstations (e.g., PC).

Specifically, the following two predictions were made:

H2a. A respondent's evaluation of an intelligent workstation will be *more positive* than that of a main-frame terminal.

H2b. Communication possibilities are believed to be *better* with an intelligent workstation than with a main-frame terminal.

Hypothesis 3. Based on the literature cited earlier, an attempt was made to compare male and female respondents in similar positions, to record differences in beliefs held. It was predicted that:

H3a. Female respondents will evaluate computers *less favorably* than their male colleagues.

Previous studies had reported mixed results about gender differences and type of computers used (e.g., *Gattiker, Gutek*, and *Berger, in press*); nevertheless, these possible effects will also be investigated further here.

Method

Design and Subjects

A stratified sample of twenty-eight employers was asked to participate in a survey of their personnel's "computer beliefs and perception of career success." Ten companies were selected at random from the *Globe and Mail* annual roster (1985) of Canada's largest organizations, twelve employers were medium-sized firms from Western Canada, three were educational institutions, and three were government agencies. Except for the stipulation of their locale (Western Canada), organizations were recruited randomly within each classification. Organizational type was not a variable of interest here. The educational and government institutions were included because they brought potentially different organizational cultures and constructs of effectiveness to the sample, thus allowing more reliable generalizations from the findings (cf. *Blalock* 1984, chap. 4).

All employers were asked to select three to six managers (female and male) and an equal number of support personnel from a variety of departments and distribute a questionnaire to these individuals. In order to avoid influencing selection decisions, organizations themselves determined whom they considered to be "managers." Surveys were returned directly to the researchers. Male respondents were matched to their female counterparts, based on similarities of job duties and responsibilities. Thus possible effects on individual perceptions of career success and computer beliefs due to differences in tasks were avoided (e.g., *Gutek* and *Larwood*, in press).

Of the 380 people asked to participate in the study, 340 agreed and 306 responses were ultimately received (90 percent). Respondents included both sexes (about 66 percent were female), and approximately 65 percent were married. Of 306 participants, 196 (64 percent) were computer users, while the remaining 110 indicated using other primary technologies, such as the telephone or typewriter, most of the time. The analyses performed for this study are based on the sample of 196 computer users, since only those answered the questions pertaining to computers.

Instrument

Respondents completed an anonymous questionnaire assessing their present position in terms of success as well as their beliefs about the technology available to them. The survey instrument was divided into three segments, with the first eliciting information about perception of success. Questions concerned global success, salary, and income as compared to peers within one's own company and beyond. Sample statements were: "I am respected by my peers," and "I am earning enough to pay my bills," with each rated on a five-point scale, ranging from (1) "agree completely," to (5) "disagree completely." Of the 38 items, 25 had been developed for a recent study (*Gattiker* 1985), while the remainder were added for this new survey.

The second part was related to one's beliefs about computer-based technology; specifically, how it supports individuals at work, whether they like using their computers, and if such use makes them more effective. The same five-point scale was provided. Of 27 items, 16 had already been developed by *Gattiker, Gutek* and *Berger* (in press) while 11 more were newly added. Additional questions about one's type of technology and the percentage of time spent using it were also included. The final section of the questionnaire concerned demographics, asking about annual income, educational background, job title, and the like.

Perception of Career Success and Computer Beliefs

Dependent variables. In order to simplify the data set, a factor analysis was conducted on the 38 success items described previously. Scales were constructed by calculating mean-scores on each factor. Those scales were then used as the variables in each of the subsequent analyses.

Predictor set. The independent predictor sets were the items measuring computer beliefs. Identical procedures as previously described were used to find the independent variables. Scales were constructed in the same way as before. The predictor scales were taken as independent variables in multiple regression analyses to determine the amount of population variance explained in perception of career success.

Demographics and Computer Beliefs

Dependent variables. These variables were the four factors extracted from the 27 items measuring computer beliefs. Scores on the factors were used in mul-

tivariate and univariate analyses of variance to determine any differences in employee beliefs about their computer-based technology.
Independent variables. The variables sex and hierarchical level in the organization (manager or support personnel) were supposed to distinguish between respondents' computer beliefs. Computer type (intelligent workstation versus main-frame terminal) was also included in these analyses. Univariate and multivariate analyses of variance were done to test for possible differences.

Results

Factors in Perception of Career Success and Computer Beliefs

To obtain the independent factors, orthogonal varimax rotations and reliability analyses were done with the 38 items measuring perception of career success. All 38 items were retained to define four factors which are labelled as follows: (1) job success, (2) financial success, (3) interpersonal success, and (4) general career success. The same analyses were performed with the 27 items measuring computer beliefs. Again, all items were retained to define the following four factors: (1) quality of job life, (2) work effectiveness, (3) communication, and (4) control.

Table 1: Items Used to Define Four Factors: Self-Perceived Disposition of Career Success

Factor	Items	Variance Factor Loadings	Explained per Factor	Item-Total Correlation	Cronbach's Alpha
	I am				
1 Job Success	receiving positive feedback about my performance from all qtrs.	.46		.50	
	offered opportunities for further education by my employer	.50		.46	
	pleased with the promotions I have received so far	.66		.63	
	having enough responsibility on my job	.64		.45	
	in a job which offers promotional opportunities	.58		.52	
	reaching my career goals within the time frame I set	.64		.62	
	fully backed by management in my work	.61		.58	
	going to reach my career goals	.43		.48	

Factor	Items	Variance Factor Loadings	Explained per Factor	Item-Total Correlation	Cronbach's Alpha
	in a job which offers me a chance to learn new skills	.74		.60	
	most happy when I am at work	.46		.40	
	offered challenges at my work	.73		.61	
	having my superior's confidence	.53		.54	
	in a position to do mostly work which I really like	.65		.54	
	in a position to set my own goals	.60		.62	
	enjoying the challenging goals I have on my current job	.73		.66	
	praised often by my superior(s)	.55		.55	
	dedicated to my work	.53		.41	
	offered opportunities for promotion by my employer	.62	26.4	.61	.90
2 Financial Success	receiving fair compensation compared to my peers	.60		.56	
	drawing a high income compared to my peers	.50		.35	
	earning as much as I think my work is worth	.77		.68	
	earning enough to pay my bills	.61		.50	
	obtaining a salary which supports my current lifestyle	.72		.62	
	paid well when compared to similar jobs in other companies	.74	6.1	.64	.80
3 Interpersonal Success	getting positive performance feedback from my peers	.59		.50	
	setting my own timeframe for career goals	.40		.43	
	often doing something with my peers outside of work	.55		.34	
	often asked for advice on private matters by my peers	.66		.50	
	frequently getting feedback from my peers about my performance	.65		.52	
	consulted often to advise a colleague on a job matter	.58	5.0	.46	.72
4 General Career Success	respected by my peers	.51		.52	
	happy with my private life	.71		.62	
	accepted by my peers	.65		.61	
	enjoying my non-work activities	.73		.65	
	satisfied with my life overall	.65		.63	
	having the confidence of my peers	.61		.58	
	enjoying spending my spare time with friends	.48		.39	

Factor	Items	Variance Factor Loadings	Explained per Factor	Item-Total Correlation	Cronbach's Alpha
	enjoying a happy family life (spouse/ partner, children, etc.)	.64	8.2	.56	.84
Total Variance explained			45.7		

Note: The above factors were obtained using principal components analysis. Orthogonal varimax rotations were performed on the data. Only loadings greater than .30 were statistically significant ($p < .001$) according to the Burt-Banks criterion (*Child* 1970).

Table 2: Items Used to Define Four Factors: Technology Disposition

Factor	Items	Factor Loadings	Variance Explained per Factor	Item-Total Correlation	Cronbach's Alpha
1 Quality of Job Life	This equipment is fun to use	.79		.71	
	Using this piece of equipment makes my work more interesting	.84		.76	
	I enjoy using this piece of equipment	.84		.81	
	This piece of equipment enables me to do interesting tasks at work	.72		.55	
	The use of this piece of equipment makes my work more enjoyable	.82		.82	
	I like doing my work with the help of this equipment	.76		.78	
	My work wouldn't be as much fun if I could not use the equipment	.64	25.1	.61	.90
2 Work Effectiveness	At my work I depend a great deal on this equipment	.47		.38	
	This piece of equipment enables me to do my job more effectively	.69		.68	
	This piece of equipment makes my work easier	.66		.63	
	This piece of equipment supports me in my work	.56		.53	
	I am more effective in work with this equipment than I would be without it	.68		.65	
	Using this equipment makes me more productive	.73		.72	
	This piece of equipment enables me to do my work faster	.74		.70	
	This piece of equipment enables me to do my job more thoroughly	.63	12.4	.70	.87

Factor	Items	Factor Loadings	Variance Explained per Factor	Item-Total Correlation	Cronbach's Alpha
3 Communication	This piece of equipment facilitates communication among people in the organization	.77		.56	
	This piece of equip. facilitates communication with people out of org.	.74		.55	
	This piece of equip. allows me to transmit information to somebody else	.68		.52	
	I prefer a face-to-face meeting over using this equipment for important matters	.34		.15	
	The use of this equipment has improved communication beyond the organization (compared to previous methods)	.74	7.9	.58	.71
4 Control	I feel this piece of equipment controls my behavior at work	.60		.40	
	This piece of equipment makes my work more demanding	.50		.29	
	My productivity is controlled by this equipment	.63		.41	
	Using this equipment limits my ability to move around	.54		.31	
	Generally, I prefer to communicate with equip. rather than face-to-face	.48		.31	
	If this equipment is out of order, I cannot do my work	.53		.28	
	It is important to me that this equipment be in use throughout the day	.53	6.8	.30	.61
Total Variance Explained			52.2		

Note: The above factors were obtained using principal components analysis. Orthogonal varimax rotations were performed on the data. Only loadings greater than .30 were statistically significant ($p < .001$) according to the Burt-Banks criterion (*Child* 1970).

Predicting Perception of Career Success

Hypothesis 1a. This hypothesis stated that the perceived influence of the technology upon quality of job life and work effectiveness would predict the largest part of the population variance in job success and general career success. To obtain the unique adjusted R^2, the composite measuring quality of job life was entered into the regression by itself. Work effectiveness, communication, and control were also subjected to separate regression runs.

The quality of job life measure accounted for a significant amount of the population variance (10 percent) when predicting an individual's perception of job success (see Table 3). Within this factor, we examined Pearson's r to determine the direction of the factor contribution, as suggested by *Cohen* and *Cohen* (1983, chap. 3). Quality of job life correlated positively ($p<.001$, by a two-tail test of Pearson's r) with job success. This indicates that respondents who enjoy their work feel that they have greater job success.

Table 3: Disposition of Career Success: R^2 by Each Predictor Set

Predictor Set	Number of Variables	Job Success	General Career Success	Financial Success	Inter-Personal Success
Quality of Job Life	1	.10****	.05****	.01	.02
Work Effectiveness	1	.08****	.07****	.01	.02*
Communication	1	.04***	.00	.00	.01*
Control	1	.00	.03***	.00	.00
Cumulative Adjusted R^2		.13***	.12****	.00	.04

Note: Adjusted R^2 is an estimate of the population R^2 adjusted for the number of predictors (*Cohen* and *Cohen* 1983: 105–107) with the multiple regression values actually obtained. The unique contribution of the adjusted R^2 of work enjoyment, work effectiveness, communication and control was obtained with separate regression runs. To obtain the cumulative adjusted R^2, all predictor variables were entered together in the regression at the same step/time. Job success, general career success, financial success and interpersonal success represent the factors obtained.

*$p < .05$
***$p < .001$
****$p < .0001$

Work effectiveness also accounted for a significant part of the population variance in job success (8 percent), showing a positive correlation ($p<.001$, by a two-tail test of Pearson's r). One interpretation could be that a respondent who perceives the computer as aiding his/her work effectiveness tends to feel more successful on the job.

Communication explained only 4 percent of the population variance in job success. The two composites correlated positively ($p<.01$, by a two-tail test of Pearson's r), thus the workers who think the technology helps them to communicate better within and beyond the organization may also perceive greater job success.

Quality of job life accounted for 5 percent of the population variance when predicting an individual's perception of general career success. These two composites correlated positively ($p<.001$, by a two-tail test of Pearson's r), meaning that those individuals who enjoy their work also feel more successful.

Work effectiveness explained 7 percent of the population variance when predicting general career success. Again, these two composites correlated positively ($p<.001$, by a two-tail test of Pearson's r), showing that a respondent who perceives the computer as aiding his/her work effectiveness also believes he/she has greater general career success.

Control accounted for only 3 percent of the population variance when predicting an individual's perception of general career success. The correlation with general career success was negative ($p<.01$, by a two-tail test of Pearson's r), suggesting that those employees who feel computer technology controls their behaviour at work do not feel successful in their overall careers.

Based on these results, *Hypothesis 1a* seems *confirmed*. It suggests that a person's beliefs about computers and their effects on quality of job life and work effectiveness relate to his/her perception of job success and general career success. Thus, having a positive belief about working with computers correlates with a positive subjective perception of job and general career success.

Hypothesis 1b. This hypothesis held that all predictor variables together would explain a significant part of the population variance in job and general career success. Results show, however, that quality of job life and communication, by themselves, already account for nearly the same amount of the population variance explained in prediction of job success as do all four predictors combined (cumulative adjusted $R^2 = .13$) (cf. Table 3). The cumulative R^2 also indicates a net suppression, which is explained by the fact that all scales are positively correlated (*Cohen* and *Cohen* 1983: 87–92).

Table 3 further demonstrates that quality of job life and work effectiveness, by themselves, account for nearly the same amount of the population variance explained in the prediction of general career success as do all four factors combined (cumulative adjusted $R^2 = .12$). Again, all r's are positively correlated, leading to a net suppression (*Cohen* and *Cohen* 1983: 87–92).

Generally, the results listed in Table 3 *support Hypothesis 1b*. In other words, a person's overall computer beliefs and their effect on his/her work situation help explain why certain individuals may perceive themselves to have more success on the job and in their career than others.

Types of Computer-Based Office Technology

Two types of computer-based technology were examined in these analyses: main-frame terminals and intelligent workstations (personal computers and word processors). Survey respondents were grouped according to the type of computer used most often. Multivariate and univariate analyses of variance were used to compare the scores of each of the groups on the four factors outlined previously.

Hypothesis 2a. Respondents were expected to differ in their evaluation of a technology, depending upon its type. The results of the analyses for the two types of computer are shown in Table 4.

Table 4: Multivariate Analysis of Variance for Beliefs About Computers

Source	df	Multivariative Tests of Pillai's *V* *F*	(df)	Univariate *F*-tests Quality of Job Life	Work Effectiveness	Communication	Control
C (type of Computer)[1]	1	3.12*	(4,172)	4.60*	.55	4.40*	4.26*
C with effects of M removed [4]		2.51*		3.40	.19	3.73*	3.61*
C with effect of S removed [5]		2.95*		3.29	.36	6.87*	1.20
C with effects of M & S removed[6]		2.62*		2.85	.20	6.15*	1.23
M (Hierarchical Level)[2]	1	1.31	(4,172)	2.36	1.98	.92	.89
M with effects of C & S removed		1.00		.94	1.56	1.18	.03
S (Sex)[3]	1	3.37**	(4,172)	1.58	.27	.94	8.14**
S with effects of C & M removed		3.51**		.05	.01	4.33*	5.14*

Note: Multivariate tests compare the two groups (e. g., managers vs support personnel) on all *four factors simultaneously,* using Pillai's *V* as calculated by SPSSX MANOVA, whereas univariate tests compare the groups on *one factor at a time* only.

1) C = type of computer used (mainframe terminal versus intelligent workstation)
2) M = hierarchical level of respondent (manager versus support personnel)
3) S = gender of respondent (female versus male)
4) Looking at computer effects having removed possible effects by the person's hierarchical level in the organization.
5) Looking at computer effects having removed possible effects by the person's gender.
6) Looking at computer effects having removed *both* possible effects of a person's gender and hierarchical level on his/her responses.

* $p < .05$
** $p < .01$

The multivariate test (*F* tests of Pillai's *V* from SPSSX Manova) showed reliable differences (F=2.62, p<05) between main-frame terminal users and intelligent workstation users on the pattern of their scores on the four factors. The means indicate that computer end-users who work mostly with intelligent workstations have more positive beliefs about computers than their peers who utilize main-frame terminals ($p<.05$, by a two-tail t-test of the scale means for both groups). Thus, *Hypothesis 2a* appears *supported*.

Hypothesis 2b. It was claimed that both groups of computer users would differ in their evaluation of the technology's influence upon communication within and beyond the organization. Univariate tests using scores on the individual factors indeed revealed group differences for communication (cf. Table 4, fourth row – "C with effects of M & S removed").Clearly, the communication factor distinguishes among individuals as to the type of computer used. The means derived from the scales indicate that individuals working with intelligent workstations felt that communication improved significantly as opposed to main-frame terminal users who felt that it did not ($p<.01$, by a two-tail t-test between the scale means of the two groups). End-users working with intelligent workstations (e.g., a PC) have a positive set of beliefs which indicates that the technology may help their communication with others both within and beyond the organization. Main-frame users, however, feel less positive in this respect. Based on the above, *Hypothesis 2b* seems *confirmed*.

Beliefs About Computers and Demographics

Individual evaluations of computer-based technology were examined according to gender and hierarchical level (managers versus support personnel). Again, multivariate and univariate analyses of variance were used to compare the scores of each of the groups on the four factors.

Hypothesis 3a. Hypothesis 3a suggested that individual evaluations of computer-based technology would differ according to gender. As Table 4 demonstrates (cf. last row – "S with effects of C & M removed"), the multivariate test (F test of Pillai's $V = 3.51$) shows a highly reliable difference ($p<.01$) between the men and women in the pattern of their scores on the four factors. The univariate tests using scores on the individual factors revealed group differences in the analyses for both communication and control. There were no reliable differences, however, between a person's gender and his/her perceived quality of job life and work effectiveness.

These results *confirm Hypothesis 3a*, indicating that gender can help distinguish end-user beliefs about computers. Specifically, women had less positive beliefs about computers and their effects upon communication than men had. Women, when compared to men, also felt that computers exercised more control over their work. Since male respondents were matched to females in terms of similar job tasks, these results become even more significant.

Type of computer and gender of respondent. Based on earlier research, a test was indicated to see if gender would account for differences of opinion between the two user groups. However, previous research is scarce, making hypothesis testing inappropriate.

If gender and type of computer are analyzed simultaneously with a multivariate test of Pillai's V (F =2.51), the respondents differ reliably ($p<.05$) in how they evaluate their computers (cf. Table 4). The univariate tests show that the respondents differ in their evaluation of the communication and control factors. The means derived from the scales indicate that women working with main-frame terminals perceived increased control by their technology ($p<.01$, by a two-tail t-test between the scale means of the two groups). They did not differ statistically significantly from male respondents in their perception of control exercised by intelligent workstations. The means also show that female respondents differ in their evaluation of computer-aided communication. Women perceived less improvement in communication possibilities from intelligent workstations than men did ($p<.01$, by a two-tail t-test of the two scale means). Yet, when assessing communication with main-frame terminals, men and women did not differ significantly.

Type of computer and hierarchical level. As Table 4 illustrates, position in an organizational hierarchy is no indicator for differentiation between respondents. Individuals participating in this survey, therefore, did not differ along hierarchical lines (managers versus support personnel) in their evaluation of computer-based technology.

Discussion

The primary research purpose here was to examine perceptions of career success and employee beliefs about computers. These are likely to affect the user's quality of work life, productivity, and absenteeism (*Kahn* 1981). Furthermore, such beliefs, together with subjective perceptions of career success, have been identified as crucial dimensions when determining the effective use of computer technology in organizations (e.g., *Carter* 1984, *Pava* 1983).

Another major focus of this study was to see if computer beliefs would differ according to demographic factors such as gender, hierarchical level in organization, and type of computer used in one's work. Respondents to this survey had worked an average of four years (lowest 2.5) with the technology. Any novelty effect would have worn off after that length of time. Previous research has indicated that individual beliefs tend to be very stable in the long run, affording the results reported here further importance (*Staw, Bell*, and *Clausen* 1986).

Perceptions of Career Success and Beliefs About Computers

The present data suggest that at least two dimensions (i.e. quality of job life and work effectiveness) can be isolated as parts of subjective perception of career success. A most intriguing result is that financial and interpersonal success could not be explained by the two dimensions, except that there is a positive correlation between work effectiveness and interpersonal success. This strongly supports other conceptual research which contends that computerization generally does not change remuneration levels nor the patterns of interaction with peers and superiors in an organization (*Salzman* and *Mirvis* 1985). The results, as they pertain to general career success and job success, will be discussed in the sections below.

Job success. According to this study, a person's beliefs about technology predict his/her perception of job success. The only item which did not predict was control. This is interesting, since job attitude research revolves around job enrichment (*Hackman* and *Oldham* 1980). Enrichment attempts to improve the individual's feeling of autonomy based on the job's responsibilities. The data here, however, indicate that technological change and working with computers in the office *may not* be perceived to influence one's perception of job success in exercising control over work habits, such as controlling speed or making it impossible to perform one's job without the technology. The data in Table 3 also show that quality of job life is the most important single predictor of job success, while communication seems to be of only minor importance.

The results indicate that employee beliefs about computers, and especially the technology's effect on quality of job life, work effectiveness, and communication, predict one's perception of job success significantly. *Brousseau* (1983) suggested that a more ideal job-person fit will be achieved if an individual views technology favorably, because such an attitude will influence his/her perception of career success positively. His contention is supported by the results of this research project. It would appear then, that if technological developments at work are agreeable, the person's subjective assessment of job success improves.

General career success. Table 3 further indicates that communication is not a significant predictor of perceptions of career success, whereas control is. An employee's belief about a computer's impact on his/her work effectiveness is the single most important predictor, whereas control is only of minor importance. Neither predictor is of great significance, though. Overall, only 12 percent of a person's perception of career success, which includes general related areas, is influenced by his/her belief about technology at the workplace. This, in itself, is an important discovery.

Earlier work has indicated that a person's assessment of career success may go beyond job-related tasks. This study represents a step toward understanding how technological change may influence a person's subjective assessment of general career success. The literature on quality of work life has indicated that experiences at work could influence a person's overall quality of life (e.g., *Podgorecki* 1981). Nevertheless, very rarely have researchers attempted to substantiate such conceptual claims with their gathered data. To some extent, this study appears to confirm that experiences at work can influence a person's overall quality of life.

Conclusion. As the cumulative adjusted R^2 shows, neither perceived ease of communication with, nor control by, computers add significantly to the variance accounted for by work effectiveness and quality of job life. This result now stands opposite a claim by other researchers that computerization will change communication as well as control exerted over employees (e.g., *Zuboff* 1982). The data here do not necessarily contradict such earlier work; rather, they point out objective changes in these areas.

Computer Beliefs and Demographics

Research investigating possible differences in individual's beliefs about computer-based technology due to demographics is limited. The data in this project, however, indicate that differences can be found even though objective task differences have purposely been eliminated. As Table 4 shows, hierarchical differences in respondents' beliefs about technology could not be found.

Type of computer used. The results show that people assess computers differently depending on whether they use intelligent workstations or main-frame terminals. In contrast to other work (e.g., *Gattiker, Gutek*, and *Berger*, in press), respondents did not differ in their evaluation of quality of job life. Instead, users varied in how they assessed communication and control. The data indicate, however, that if the effects of gender are removed, differences in one's belief about control according to type of computer used cannot be distinguished. For communication, though, the data shows that intelligent workstation users feel that computers affect their communication possibilities more positively when compared to their peers who work with main-frame terminals.

Gender. In this study, the sexes differed in how they assessed the computer's possible influence on *communication and control*. The data revealed that men, more than women, felt that intelligent workstations *improved* communication within and beyond the organization, while the sexes did not differ in their assessment of communication using a main-frame terminal. The data further showed that female respondents were more likely to believe that communication would be affected negatively by their computer. Thus, communication within and beyond the organization, as well as communication with one's peers, was not necessarily perceived as having improved with the introduction of a computer.

The control factor, on the other hand, helped to distinguish respondents according to gender. Women felt that they were dependent upon the computer to perform their tasks, and felt somewhat controlled in their work speed. Also, women felt *less* in control when working with a main-frame terminal, but the perceptions of control when working with an intelligent workstation were about equal for men and women. One must remember that a random stratified sample was used, including both women and men from similar hierarchical levels and job categories, making this finding even more interesting.

There are several possible explanations for the results of this research. The most obvious one could be that traditionally female clerical jobs using main-frame terminals and/or typewriters may be changed to using intelligent workstations. *Morgall* (1983) has already reported that peers tend to avoid communicating verbally to an individual when he/she is working on a computer. This could, in part, explain why women felt less positive about the communication possibilities of intelligent workstations. Working with a typewriter in the past may have allowed for more informal verbal communication than they feel is currently possible.

An explanation for the control effect could be that, in most female-dominated clerical jobs making use of a main-frame terminal, quantitative measures are used to assess performance (e.g., claims recorded per hour) (*Gutek* and *Larwood*, in press). Even though the women in this sample did not hold such positions, they might feel controlled, nevertheless. Also of importance is the result that women and men *did not differ* in how they felt about computer-based technology's effect on their quality of job life and work effectiveness, neither for intelligent workstations nor main-frame terminals.

Conclusion. These results showed that individual beliefs about technology differ in several important factors. The most significant may be that it seems intelligent workstations are perceived *more positively* than main-frame terminals. Additionally, respondents differed according to their gender. Female respondents were less positive about the computer than their male counterparts when it came to both control and communication. In particular, women feel less positive about computers than men do for both intelligent workstations and main-frame terminals. The results indicate that it is possible to distinguish groups working with computers by measuring beliefs about the technology.

In sum, the effects of computerization on employees' beliefs are complex. A simple examination of differences according to sex would be misleading; therefore, this study made sure to use respondents of both sexes occupying similar positions. Besides well documented objective effects (e.g., *Gutek* and *Larwood*, in press), the differences recorded here indicate that subjective effects are also apparent.

Implications for Management and Future Research

A thorough understanding of the effective use of any computer-based office technology is substantially reduced if employee beliefs about technology are ignored (*Blackler* and *Brown* 1985). Organizations must consider the technology's influence on individual perceptions of career success if an improved job-person fit is to be achieved. This study presents strong evidence for placing future research on computer-based technology within a correspondingly larger context.

Practical implications. Some researchers have concluded from their data that women are more affected by, and respond differently to, computerization because of the jobs and occupations they hold (*Form* and *McMillen* 1982). The variances reported here, however, might simply indicate a gender difference in work outlook and beliefs about computers (*Morgall* 1983). Computerization might also affect managers and support personnel differently, since their responsibilities vary a great deal (e.g., *Gattiker, Gutek* and *Berger*, in press). There are longitudinal studies about jobs, job design, and skills which indicate that a self-selection process is taking place (e.g., *Spenner* 1983). People choose occupations and jobs which fit their individual needs in such areas as job complexity and decision-making (*Schein* 1978, chap. 8). For instance, programming at a terminal may appear preferable to endless hours of data entry to an outsider (cf. *Gutek* 1983, *Mankin, Bikson,* and *Gutek* 1982), yet those in such positions probably would not want to trade places. Therefore, objective differences may not transcend the individuals' subjective assessments of career success and computers.

In terms of practical issues, the most obvious implication in these findings concerns the type of computer-based technology to be used. The data in this study and others (e.g., *Gattiker, Gutek,* and *Berger,* in press) indicate that individual's beliefs about intelligent workstations are *far more positive* than they are for main-frame terminals. Thus, increased acceptance and effective use of computers may be greatly facilitated by installing intelligent workstations. This is an important finding for managers who must decide about their firm's future strategies in the area of office automation.

For managers, the results obtained present new confirmation of the notion that computerization of work does affect certain aspects of an employee's perception of career success. These effects, however, are relatively insignificant for financial and interpersonal success. Nevertheless, there are implications for career development. It seems that individuals are affected by their beliefs about computers in how they perceive job success and general career success. Apparently, intelligent workstations may facilitate the individual's positive beliefs about computers, and his/her perception of career success.

These results, together with other studies (e.g., *Gattiker, Gutek,* and *Berger,* in press), suggest that productivity increases via computerization might best be achieved by providing employees with intelligent workstations which possess main-frame communication abilities. Since the respondents felt most comfortable with intelligent workstations, it is safe to assume that their productivity levels will ultimately be higher than if they worked with main-frame terminals only (*Carter* 1984, *Kahn* 1981). Women appear more afraid of computers than do their male colleagues, which may explain their less positive beliefs reported in this study. One possible strategy to overcome fears and foster acceptance might be additional education and information for female workers about computers and their effect on their work, as suggested by some researchers (cf. *Dierkes* and *Von Thienen* 1984).

Theoretical implications. Even though we may not know exactly how an employee's computer beliefs and his/her perception of career success have developed, the finding that the former can help explain the latter has some important implications. In terms of job design, it suggests that we may have erred too much towards situational determinism. *Staw, Bell*, and *Clausen* (1986) argued that we ought to increase our knowledge about moderators of beliefs, which in this study is one's type of computer.

Sorting out the interaction of beliefs and job characteristics is likely to be a complex research task, as already illustrated by some recent studies in occupational sociology and management. For instance, *Gattiker* (1987) found that a positive assessment of subjective career success and computers did not necessarily lead to increased commitment or lower levels of role conflict and ambiguity. *Kohn* et al. (1983) found that intellectual flexibility is affected by a person's job and can create reciprocal causal loops that are difficult to understand and study. Future research efforts would do well to combine aspects of three traditions: psychology, sociology, and information systems management.

An attempt has been made here to expand the existing research on computer-based technology by studying user beliefs. The relationship between those beliefs and perception of career success shows that job success and overall career success are highly correlated with the individual's evaluation of computer impact upon work effectiveness and quality of job life. Additionally, gender differences among computer users have been shown and tested. As this direction is pursued further, new theory and practical insights will be gained which may suggest not only revisions to managerial practice, but also to our general approach to office automation.

Notes

1 Perceptions are the cognitive representations of attributes that reflect the psychological meaning and significance of these attributes to individuals (i.e., the perceived career success in areas such as income, peer relationships, job success, and general career success).

2 In this paper, beliefs about computer-mediated work describe a person's cognitive and affective states about the effect of the computer on his/her job situation and related tasks,

i.e. his/her attitude (*Gutek* 1983). An attitude is generally seen as a disposition to respond in a *favorable or unfavorable manner* to an object or occurrence (*Oskamp* 1977: 2–12).

3 The term *computer-based technology*, as used in this paper, encompasses, but is not limited to, main-frame terminals and intelligent workstations (e.g., PC's).

References

Bandura, A. (1982): The psychology of chance encounters and life paths, *American Psychologist*, 37: 747–755.

Blackler, F. and *Brown, C.* (1985): Evaluation and the impact of job information technologies on people in organizations, *Human Relations*, 38: 213–231.

Blalock, H. M. (1984): *Basic dilemmas in the social sciences,* Beverly Hills, CA: Sage Publications.

Brousseau, K. R. (1983): Toward a dynamic model of job-person relationships: Findings, research questions, and implications for work system design, *Academy of Management Review*, 8: 33–45.

Carter, N. M. (1984): Computerization as a predominate technology: Its influence on the structure of newspaper organizations, *Academy of Management Journal*, 27: 247–270.

Child, D. (1970): *The essentials of factor analysis*, New York: Holt, Rinehart & Winston.

Cohen, J. and *Cohen, P.* (1983): *Applied multiple regression/correlation analyses for the behavioral sciences* (2nd ed.), Hillsdale, NJ: Lawrence Erlbaum Associates.

Demby, E. H. (1985): The new requirement for the researcher: Interactivity with the personal computer, *European Research*, 13: 26–28.

Dierkes, M. and *Von Thienen, V.* (1984): *Kein Ende der Akzeptanzschwierigkeiten moderner Technik? Zum Zusammenhang der informationstechnischen Entwicklung und ihrer Akzeptanz.* (Report No. P84–2), Berlin: Science Center Berlin.

Doswell, A. (1983): *Office automation*, New York: John Wiley and Sons.

Fishbein, M. (Ed.) (1967): *Readings in attitude theory and measurement*, New York: John Wiley and Sons.

Form, W. and *McMillen, D. B.* (1982): Women, men, and machines, *Work and Occupations*, 10: 147–178.

Gattiker, U. E. (1984): Managing computer-based office information technology: A process model for management, in: *H. W. Hendrick* and *O. Brown, Jr.* (Eds.), *Human factors in organizational design*, pp. 395–403, Amsterdam: Elsevier Science.

Gattiker, U. E. (1985): Human resources utilization and career management, *Proceeding of the National Meeting of the Association of Human Resources Management and Organizational Behavior*, pp. 458–471, New York: Maximilian Press.

Gattiker, U. E. (1987): A dispositional approach to office automation: Testing a causal model. (Manuscript submitted for publication.)

Gattiker, U. E. (in press): Organizational change: A typology for strategic human resource management, *Behaviour and Information Technology*.

Gattiker, U. E. and *Larwood, L.* (1986): Subjective career success: A study of managers and support personnel, *Journal of Business and Psychology*, 1: 78–94.

Gattiker, U. E. and *Nelligan, T. W.* (1988): Computerized offices in Canada and the United States: Investigating dispositional similarities and differences, *Journal of Organizational Behavior*, 9: 11–96.

Gattiker, U. E., Gutek, B. A., and *Berger, D. E.* (in press): Office technology and employee attitudes, *Social Science Computer Review.*

Gutek, B. A. (1983): Women's work in the office of the future, in: *J. Zimmerman* (Ed.), *The technological woman*, New York: Praeger.

Gutek, B. A. and *Larwood, L.* (in press): Information technology and working women in the USA, in: *M. J. Davidson* and *D. L. Cooper* (Eds.), *Women and technology,* Chichester, UK: Wiley & Sons.

Hackman, J. R. and *Oldham, G. R.* (1980): *Work redesign*, Reading, MA: Addison Wesley.

Kahn, R. L (1981): Work and health: Some psychosocial effects of advanced technology, in: *B. Gardell* and *G. Johansson* (Eds.), *Working life,* pp. 17–37, Chichester, UK: Wiley and Sons.

Kiesler, S., Zubrow, D., Moses, A., and *Geller, V.* (1985): Affect in computer mediated communication: An experiment in sychronous terminal-to-terminal discussion, *Human-Computer Interaction*, 1: 77–104.

Kling, R. (1978): Information systems and policy making: Computer technology and organizational arrangements, *Telecommunication Policy,* 2: 22–32.

Kling, R. and *Iacono, S.* (1984): Computing as an occasion for social control, *Journal of Social Issues*, 40: 77–96.

Kohn, M. L. and *Schooler, C.,* with the collaboration of *Miller, J., Miller, K. A., Schoenbach, C.,* and *Schoenberg, R.* (1983): *Work and personality: An inquiry into the impact of social stratification,* Norwood, NJ: Ablex.

Korman, A. K. (1980): *Career success/Personal failure,* Englewood Cliffs, NJ: Prentice-Hall.

Mankin, D., Bikson, T. K., and *Gutek, B. A.* (1982): The office of the future: Alternatives and choices, *The Futurist*, 16: 33–37.

Megaw, E. D. and *Lloyd, E. J.* (Eds.) (1984): Human-computer interaction: A special issue of *Ergonomics Abstracts*, 16 (1).

Morgall, J. (1983): Typing our way to freedom. Is it true that office technology can liberate women? *Behaviour and Information Technology*, 2: 215–226.

Oskamp, S. (1977): *Attitudes and Opinions,* Englewood Cliffs, NJ: Prentice-Hall.

Panko, R. R. (1984): Office work, *Office: Technology and People,* 2: 205-238.

Pava, C. H. P. (1983): *Managing new office technology*, New York: Free Press.

Podgorecki, A. (1981): Pathology of institutions and organizations versus 'quality of working life,' in: *B. Gardell* and *G. Johansson* (Eds.), *Working Life*, pp. 307–312, Chichester, UK: Wiley and Sons.

Romney, A. K., Smith, T., Freeman, H. E., Kagan, J., and *Klein, R. E.* (1979): Concepts of success and failure, *Social Science Research*, 8: 302–326.

Rosenbaum, J. E. (1985): Persistence and change in pay inequalities: Implications for job evaluation and comparable worth, in: *L. Larwood, B. A. Gutek,* and *A. Stromberg* (Eds.), *Women and work*, (Vol. 1), pp. 115–140, Beverly Hills, CA: Sage Publications.

Salzman, H. and *Mirvis, P.* (1985): *The workforce transition to new computer technologies: Changes in skills and quality of work life* (Tech. Rep. 85–8), Boston: Boston University, Center for Applied Social Science.

Schein, E. H. (1978): *Career dynamics: Matching individual and organizational needs*, Reading, MA: Addison Wesley.

Spenner, K. L. (1983): Deciphering Prometheus: Temporal change in the skill level of work, *American Sociological Review*, 48: 824–837.

Staw, B. M., Bell, N. E., and *Clausen, J. A.* (1986): The dispositional approach to job attitudes: A lifetime longitudinal test, *Administrative Science Quarterly*, 31: 56–77.

Zuboff, S. (1982): New words of computer-mediated work, *Harvard Business Review,* 60: 142–152.

Chapter 9
The First-Line Supervisor and Advanced Manufacturing Technology*

John E. Ettlie

Abstract

The individual that occupies the role of the first-line supervisor continues as the "forgotten person" into the era of deployment of high technology in manufacturing. This paper suggests that the role of supervision is particularly sensitive to technological change in plants that are participating in the rapidly increasing diffusion of advanced manufacturing technology. Preliminary data are presented comparing first-line supervision with other occupational groups in domestic plants that have installed new flexible processing technology. Supervisors were found to have the most routine jobs after new, flexible production technology was installed, significantly different from managers and engineers. With no particular advantage in other task content areas such as skill variety, autonomy, and dealing with others, supervisors appear to be rather vulnerable to change stressors in coping with new technology, especially during the early stages of system startup. Implications for further research and the management of high technology plants are presented.

Introduction

We have entered an era in the study of complex organizations when the role of the supervisor has been taken for granted. Yet, many manufacturing and service sector firms are in the midst of drastic cost cutting programs that include staff reductions. In manufacturing, especially for durable goods processed in batches, the adoption of new, computer-based processing technologies is becoming widespread in response to global competitive pressure. Rather than

* Work in this area was supported in part by the National Science Foundation, the Industrial Technology Institute, and the School of Business Administration, University of Michigan. The opinions in this paper are those of the author and do not necessarily reflect the official position of the funding agencies. Assistants on this project were Mr. Calvin Chang, Ms. Stacy Reifeis and Ms. Sandy Judkins. The author gratefully acknowledges the comments of reviewers, Professor Janice Klein, and participation of plant personnel in this study.

disappearing, the role of the supervisor is changing drastically in these new production technology plants.

Recently, a report issued by the Innovative Manufacturing Council, a group of companies including the Timken Company, Eaton Corporation, and Cooper Industries, said that "first-line supervisors have the most significant impact on any productivity improvement measures," and yet, supervisors are "the weakest link in most manufacturing operations." College engineering students recruited into first-line positions often lack experience, companies typically neglect training for the first-line and the most talented managers are often reluctant to work in remote facilities. These and other problems exacerbate the impact of automation on the first-line (*Iron Age* 1986).

Modernization often creates a new occupational elite in manufacturing plants, which makes the first-line job harder. There seems to be a growing consensus that instead of the expected disappearance of the first-line job, the role has merely taken on new, more increased responsibility. With fewer people in plants and flatter structures, the link with the manufacturing floor becomes even more crucial (*Iron Age* 1986).

The purpose of this report is to examine the role of the first-line supervisor in modernizing manufacturing plants. In particular, the focus is on the job dimensions, often considered to be a function of technology, routineness, and analyzability. In addition, degree of autonomy, human relations skills, and role stress are examined across four occupational groupings. The preliminary data of this study suggest that first-line supervisors are uniquely disadvantaged during the early stages of significant plant modernization. The implications of these findings for technology management and future research directions are presented.

The First-Line Supervisor and Technology

It is no longer fashionable to consider the first-line supervisor as problematic or as a research-worthy topic. Courses for first-line supervisors and foremen are "cut and dried" and the subject has moved out of the light of interest where it once held center stage in such ground-breaking research as the Hawthorne Studies and made famous in essays characterizing this role as the "man in the middle" or inherently stressful. Most of this work was published in the 1940's and only a mild rekindling of interest in the first-line role was stimulated by the writings on technological change in the 1960's. At that time there was a recognition that "changing technology of many parts of the business throws increasing strains on the foreman" (*Dill* 1965: 1087), primarily due to such factors as 1) the decrease in freedom to schedule work; 2) general decrease in decision-making influence given up to engineering, production control, accounting, and long-range planning; 3) the decrease in number of employees supervised; and 4) increased technical skill demands of the job (*Dill* 1965: 1087).

In a well-known study of the same period, *Woodward* (1965) found a curvilinear relationship between the span of control of first-line supervisors and type of production system technology. On average, firm success followed this pattern. The *Woodward* data show that above average firms were more frequently observed in the median category of span, i.e. 23 for unit and small-batch, 49 for large-batch and mass, and 13 for process production system technology. Effective span of first-line supervisors appears to be very dependent on the type of technology of the productive unit. On the other hand, when the technology of an organization changes significantly, we know very little about the new role requirements for various occupations.

Job Satisfaction and Stress

Actually, the issue of first-line supervisors has not disappeared, but is often submerged in other materials on more general topics. For example, *Lawler* (1978) discusses at length the problems that first-line supervisors encounter in new plant startups. The more drastic the departure in the new plant from former managerial and hierarchy practices, the more difficult the transition tends to be for the foreman in the plant.

Walton and *Schlesinger* (1979) focus specifically on the problems supervisors have when work is restructured to become a more participative work system with very autonomous teams that require the supervisory role to also be changed to become more consultative. They base their conclusions on their experience with 12 manufacturing plants, nine of which restructured work during plant startup. At least one plant had startup with a new product and "substantial amounts of new technology," but it is assumed that in most of these cases, technology was taken as constant. The plants did vary on type of industry, product and basic form of production technology from continuous process to batch and assembly, in paper, food, automotive and farm equipment.

The authors report rather predictable problems, for the most part. The supervisors tend to be taken for granted in these restructuring efforts that focus on getting subordinates to take on more responsibility. As a result, "in most cases, there was relatively less satisfaction or more dissatisfaction at the first-line supervisory level than in positions above or below them in the same plant organization hierarchy."

Perhaps the most interesting finding in this study was that in cases where supervisors took a very active, "tight rein" approach initially during new plant startup, teams were most advanced in self-management after 15 months. This indicates a distinction between startup versus steady-state conditions in these new plants and the need for different behaviors at different stages of the transition process.

As would be expected, lack of understanding of new production technology leads to problems. As a result, it is unrealistic to plan for the elimination of the first-line role. Rather, the variance in these plants suggests that management needs to plan for both "an increase as well as a decrease in the amount of supervision over time" (*Walton* and *Schlesinger* 1979: 32).

Stress among first-line supervisors is rarely studied across a variety of settings, but there are exceptions. *Rosen, Aldag,* and *Joseph* (1985) investigated stress and coping behaviors in 234 "past participants of supervisory level positions" (p. 616). The authors report that "coping behaviors and social support on the job has a strong, pervasive effect on numerous outcomes [...] and [...] such variables are a function of task and demographic characteristics" (p. 617). They found that 73 percent of the variance in workplace coping is explained by task characteristics and sex, and these, in turn, "had an effect on both measures of work-related stress."

In the initial analyses of these data (*Latack, Joseph,* and *Aldag* 1985), it was found that the task characteristics that were significantly correlated with the effective coping behaviors (positive thinking and direct action) were autonomy and dealing with others. Women were significantly less likely to adopt positive thinking as a coping behavior. Further, the authors conclude that "these findings point to task and sociotechnical systems design as potentially important elements for stress management in organizations" (p. 25) vs. potentially less effective, individual focus stress management programs.

In the *Rosen* et al. (1985) LISREL analyses, it was found, contrary to the *a priori* model, that social support was revealed as part of the workplace coping construct rather than a determinant of coping behavior. Another interesting result was the absence of any "significant relationship between life stress and the coping construct" (p. 617), and the presence of residual correlation between these variables prompted the authors to suggest that, among other possibilities, an exogenous factor such as a personality variable like type A behavior might be added. It has generally been found that stress levels are elevated during the implementation period of new processing technology (e.g., *Ettlie* and *Rubenstein* 1980), so these reports and related results may be very useful in the study of supervisors and advanced manufacturing technology deployment. Shifts in task content due to technology change may require new coping behaviors for supervisors.

The Influence of Supervision on Modernization

A few brief case histories might serve to reintroduce the importance of the role of the first-line manager during modernization, and illustrate some of the key, researchable issues as well as hypotheses for testing in changing plants.

The first case is drawn from an automotive engine plant and data collected during the fall of 1985 as part of another study (*Ettlie* 1985). In this case, as in so many others, we were trying to determine how shop floor personnel are selected to participate in the modernization effort. This automotive case was for a new integrated manufacturing line dedicated to one particular set of engine components. A first-line supervisor was asked to recommend an hourly person to be trained to operate the new line. A man was suggested and trained and lasted only three weeks on the project. Follow-up with the supervisor that made the recommendation revealed that the man selected had been a difficult employee for the foreman due to an attendance problem. Because the system was to be located in another department, the foreman saw an opportunity to transfer this employee out of his supervisory area. Engineers assigned to any project are now careful when soliciting opinions from supervisors on possible operators for new systems, distinguishing suggestions from supervisors located in the areas where a system is actually being installed and suggestions from supervisors not located in these departments. Typically, first-line supervisors exert a great deal of informal influence on the selection of operative and skilled trade personnel for advanced manufacturing technology projects.

A second case covers the more advanced stages of deployment for new production technology. Here the data were being collected for the same project but a year earlier, in the fall of 1984. This is a case of a flexible manufacturing system to machine geometric parts. The issue here was the selection of the two system supervisors themselves. Both were selected for their progressive style from many applicants, even though neither had much shop experience (less than two years in one case). Both use a "walk around" style of managing the system, and neither spends any time at all in the control booth. Another interesting feature of this case was the extreme resistance of the maintenance foreman to the system. This occurred because maintenance personnel in mechanical and electrical repair were dedicated to working part-time on the system. Apparently, the maintenance foreman's authority was undermined by the scheduling of repairs and preventive maintenance on the new system. Changing structural patterns due to new system deployment may decrease the amount of influence that supervisors have over their immediate subordinates.

A third example, which again typifies many other cases, involved a new flexible manufacturing system being tracked as part of the same study, but representing training practices across several other installations. In this case, two operators, with very general job descriptions, were being trained from the pool of operators assigned to the new project for "control room" jobs, which are equivalent to system supervisor jobs. They were being trained, in part, by the first shift system supervisor. Either this indicates the careful planning for implementation and potential turnover on the flexible manufacturing system (FMS) team, the gradual merging of system related jobs as a natural consequence of the implementation strategy that uses broad job descriptions, or the increasing similarity of first-line versus area supervisor role descriptions in modernizing plants. The pervasiveness of this practice suggests that perhaps all three or even more explanations might be used to define this strategy. New technology appears to create opportunities for mobility and a new internal labor market around the first-line supervisor role.

Case history results tend to be somewhat complementary with those predicted trends in the *Hill* and *Kerr* (1984) review of the literature on the impact of computer integrated manufacturing on the first-line supervisor. The authors predict directive leadership styles giving way to participative styles and increased skills for communication organization-wide. If there is a tendency for foreman and operators on systems to have similar tasks, and even if there are fewer operators, this does not mean they will interact more with their supervisor. They may in fact be more self-sufficient and more independent than before. Our case material is divergent with the *Hill* and *Kerr* (1984: 95) prediction that "strong shop floor knowledge" will recommend an existing first-line supervisor to a new position. In the extreme, there may, in fact, be no first-line supervisor but an area supervisor instead, and he or she may be a person with engineering training and only some shop experience. It might be interesting to track what a given firm does with failed supervisors in new system deployment, and in general, attempt to understand how supervisors are selected for modernizing programs.

Case evidence presented earlier suggests first-line supervisors do have an important informal influence over selection of employees for Computer Integrated Manufacturing (CIM) training. This disagrees with the *Hill* and *Kerr* (1984: 94) prediction that supervisors will have decreased influence in that activity area. On the other hand, increased communication skills are likely to be required as the technology core is exposed. This is consistent with the *Hill* and *Kerr* prediction of an expansion of maintaining external relations activities (p. 92).

Summarizing these data and case studies, the following observations can be made about the role of the first-line supervisor in modernization und its implications for the changing task structure in the productive segment of any organization.

1. The role of the first-line supervisor appears to be especially sensitive to technological change in organizations.
2. Effective span of control of first-line supervisors seems to vary with the technology of the production unit. It is broadest for large-batch and mass production technology, but narrower for unit, small-batch and especially

process technologies. But spans may generally be decreasing due to labor cuts.
3. Shifts in task content for first-line supervisors may require new coping behaviors to deal with the stress and lower job satisfaction created by change.
4. First-line supervisors exert considerable informal influence on the selection of shop floor personnel in the deployment of advanced manufacturing technologies, and the amount of influence they have may change when new systems are actually installed.
5. New technologies in manufacturing seem to change the internal labor market and roles in a plant. First-line supervisors are at the center of many of these changes.

Research Issues

The evolving role of supervision in manufacturing plants is assumed to be a function of not only new processing technologies but the alternative structures and strategies currently being used to deploy these new technologies as well. Hypothesis development follows.

Workgroup Technology, Autonomy, and Dealing With Others

Hill and *Kerr* (1984), in their review of the literature on the first-line supervisors and computer integrated manufacturing, predict that the first-line supervisor will require both more technical as well as increased human relations and communications skills. This suggests that first-line supervisors will become more like middle managers, relating to the total organization, rather than "heavily subordinate-oriented" (p. 94).

However, the overall level of technical expertise required in an organization will increase with modernization regardless of occupation. For past generations of computer technology, there has been a general decrease in decision-making influence of the first-line role, partly due to the increased influence of engineering (*Dill* 1965: 1087). There is at least some evidence that supervisors have attempted to cope with new technology in production by trying to reduce job complexity. *Noble* (1986: 184–185) presents the case study of Ralph Kuhn, a tool and die supervisor in the value engineering department at the Rouge River Ford plants and responsible for evaluating numerical control (N/C) on the shop floor. In his role as a production supervisor, he was preoccupied with efficiency, quality, and schedules, and evaluated numerical control technology in rather objective terms, asking unheard of questions, like whether this was the simplest and cheapest way to get the job done.

In the late 1960's, Ford had spent much on developing an in-house computer language known as FORSUR (Ford Surface) but it was a complicated and time consuming way to generate tapes. Further, Kuhn felt that it created a language barrier between those that prepared the tapes and those people that used them. Therefore, in the late 1960's, he proposed a simplified shop procedure for point-to-point, vs. complicated contouring, programming on straight mill cutting applications that was very similar to the one developed by John Parsons much earlier. This method required only 20 percent of the cost and a quarter of the time needed with the computer-based FORSUR method.

After much discussion, Kuhn got permission to go ahead with his method in the tool and die shop using machinists as programmers, and the experiment ended after a year and a half when the United Auto Workers (UAW) management "was using numerical control to displace workers and to remove workers from the bargaining unit," and were insisting that workers should retain control of the programming function, much like the Kuhn experiment recommended. Ford would not yield and the FORSUR method was adopted as standard. Kuhn contended that control was more important to management than economics. N/C engineers had joined management by being skeptical and hostile to the simplified approach.

Ironically, it was during this same period, elsewhere in Ford, that an experiment with the forerunner of flexible manufacturing was taking place, in an attempt to design and install a variable mission machining system. The supplier of the system was Milacron of Cincinnati. Eventually, this concept was abandoned as well, in spite of the fact that the purpose was really the same as Kuhn's: to increase the utilization of N/C, and, hopefully, make it a more cost effective concept for production.

According to *Noble* (1986: 185), Kuhn tried another, similar experiment a few years later, this time using a record-playback method for producing N/C tapes on three-axis profile milling, similar to the job the FORSUR system was designed to solve. Kuhn was still convinced the simpler way was more cost effective. Kuhn argued that cost-saving projections made by Ford staff groups were never realized and figures were often "jockeyed" to justify staff efforts. One of Ford's suppliers, Buffalo Tool and Die Company went bankrupt, supposedly because Ford could not supply tapes using the computer-assisted method on time. Their three-axis milling machine sat idle waiting for tapes. Kuhn also tried a related experiment with an inspection machine but was forced to abandon the project, and nothing was ever published in the professional literature on the project. In 1982, Kuhn said, "For 20 years we've been on the wrong track," referring to the demise of the U.S. auto industry. The system had been unable to reduce job complexity and the autonomy of Kuhn and his fellow supervisors was, in fact, curtailed.

Hill and *Kerr* (1984) predict a decrease in autonomy for first-line supervisors with the adoption of new production technologies. This is very similar to the result obtained by *Mann* and *Hoffman* (1956) when a new and old power plant were compared. When the technical and operator jobs were greatly expanded, all occupations became more interdependent.

One of the few empirical studies done that makes explicit comparisons between occupations for the operation of a flexible manufacturing system is reported by *Gerwin* and *Blumberg* (1984). Although this really is just a case study of one installation of flexible manufacturing, normative data were included in the data tables for comparison, making this a much more useful report. Occupations included foreman, repairman, tool setters, operators, and loaders. Overall, people assigned to the operation scored lower on job satisfaction dimensions, but there were mixed findings for stress and job characteristics. As would be expected with a more integrating technology, autonomy is lower than norm group averages. Overall, some work-related stress scores were elevated, but not on the more well-known measures like role conflict and clarity.

If one were to use the Perrow model of workgroup technology, it would be equivalent to predicting fewer exceptions and more analyzability for the position. In short, these jobs became more routine (*Withey, Daft,* and *Cooper* 1983). The following hypotheses are offered for testing.

Hypothesis 1: In modernizing plants, the resultant supervisor's role is likely to be more routine, i.e., more analyzable, with fewer exceptions, than others.

Hypothesis 2: In modernizing plants, the resultant supervisor's role is likely to have less autonomy.

Hypothesis 3: In modernizing plants, the resultant supervisor's role is likely to require greater emphasis on human relations and dealing with others at all levels.

Workgroup Technology and Role Stress

There is likely to be a complex relationship between role stress and workgroup technology during modernization. Some jobs continue as complex and challenging, like engineers and skilled trades maintenance roles. Operators and first-line roles will vary by deployment strategy. Based on *McGrath's* (1970) delineation of stress factor and response capabilities, moderate levels of workgroup technology routineness and analyzability are likely to produce the lowest levels of stress. However, high levels of stress will result from jobs either too routine and analyzable or jobs rife with exceptions – the extremes in complexity.

In the *McGrath* (1970: 17) conceptual formulation of the stress paradigm, "stress occurs when there is a substantial imbalance between environmental demand and the response capability of the focal organism." Yet, there is a substantial difference between this paradigm and other approaches to stress. Most other stress concepts imply that the imbalance in the organism-environment consists of "overload" or too much demand – beyond capabilities. But this formulation also includes situations of stimulus impoverishment, and social isolation, that all suggest "stress-like effects may result from an environment that places too little demand on (or at least provides too little or too uninteresting a pattern of stimulation for) the focal organism" (p. 18). Stress causing imbalance can, therefore, result from overload and underload.

How is this pattern likely to develop for the first-line? The first-line is likely to have high role stress because the job will become routine relative to others. *Gerwin* and *Blumberg* (1984) found that supervisors had the highest role conflict after machine loaders and highest role ambiguity after operators, so it is assumed that when one combines these results, they have the highest role stress overall for the six occupational groups compared in that flexible manufacturing system operation. Unlike the findings in *Walton* and *Schlesinger* (1979), *Gerwin* and *Blumberg* (1984) report that "foremen in spite of their difficult situation tend to be relatively satisfied with four of the six factors" (p. 177) measured, including challenge, resource adequacy, promotion, and overall financial reward.

Managers and engineers are likely to experience high stress for the opposite reason: the high number of exceptions in the role of planning for the new system. Skilled trades are likely to increase somewhat or remain at elevated levels for exceptions on the job and end up at moderate levels on these measures, so their jobs are candidates for low role stress. Operators will likely group with the first-line, with jobs becoming more routine. This predicted pattern is very similar to that hypothesized and supported by *Haddad, House,* and *Jackman* (1985). Based on the above, the following hypotheses are put forward:

Hypothesis 4: There is a curvilinear relationship between role stress and routineness (analyzability and exceptions) in modernizing plants. Lowest levels of stress are for moderate levels of routine and high stress levels for very routine or non-routine jobs.

Hypothesis 5: Role stress in modernizing plants will be highest for managers and engineers, very high for first-line supervisors and for operators and lowest for skilled trades with moderate levels of routineness (analyzability and exceptions).

Method

Sample

In order to compare first-line supervisors with other occupational groups involved in significant organizational change programs, questionnaires were distributed during interviews conducted in plants installing new, advanced production technology systems. Plant cases were sampled randomly from announced purchase plans. A total of 39 domestic facilities were included. Each was evaluated for the appropriateness of questionnaire distribution.

Most of the new production technology installations were flexible manufacturing systems. That is, in 26 of these 39 plants, new multiple machine durable part manufacturing systems were installed that ran unattended at least part of the time. These systems have dedicated materials handling systems and are run using a coordinating computer control system that schedules work and keeps track of performance. Other systems included flexible assembly installations and integrated manufacturing plants. The average, initial cost of the installations studied was $ 3 million. Data were collected in the fall of 1984 and winter of 1985.

Whenever an installation project was far enough along for operators and maintenance personnel to have been identified, questionnaires were distributed. Eighteen (67 percent) of the 27 plants eligible for questionnaires mailed back at least one survey and a total of 57 (67 percent) of the 85 forms distributed were returned and were free of missing data. Given that a total of 39 plants are participating in this longitudinal study, and with only 27 of the plants actually eligible to receive questionnaires, the results obtained from these questionnaires should be considered preliminary for testing these hypotheses.

Subjects were selected jointly with the help of key interview respondents in the study. Lists of shop floor and support personnel were generated, and "representative" members of each occupational group were selected. We did not attempt to include everyone, nor was a random sample used because, with the exception of operators, there typically is only one person in each occupation. Usually one manager, one engineer, one supervisor, and one skilled trade person worked exclusively on the project. Operators were selected to represent shifts.

Scales

Four scales were used for the group comparisons and analyses. First, the two forms of the *Hackman* and *Oldham* (1980) Job Diagnostic Survey were used with subscales for skill variety, autonomy, and dealing with others. Second, role stress with the subscales for role ambiguity and role conflict was measured using the *Rizzo, House,* and *Lirtzman* (1970) instrument. This scale has been recently validated by *House, Schuler,* and *Levanoni* (1983). Third, work climate was measured using an adapted instrument from *Taylor* and *Bowers* (1969). Fourth, and lastly the *Withey* at al. (1983) scale was used to measure routineness in work unit or workgroup technology using exceptions and analyzability. Cronbach alpha reliabilities are reported in Table 1. All scales had reliabilities at or in excess of .85. The first form of the *Hackman* and *Oldham* scale had a reliability of .86, and the second scale had a .85 Cronbach alpha. The role stress scale had an alpha of .90. The work climate scale had a .92 coefficient alpha, and the workgroup technology scale had an alpha of .90.

Analyses

Two statistical analysis techniques were used to evaluate these data. The primary method used was one-way Analysis of Variance (ANOVA), using Scheffé ranges ($p = .10$) to accomplish multiple comparisons between occupational groups. A second technique was a test for curvilinearity or second order equation using regression. This F-test evaluates the significance of the increase in explained variance of entering the predictor and then the square of the predictor in a step wise regression (cf. *Namboodiri, Carter,* and *Blalock* 1975). Also, the rule of thumb of calculating the eta coefficient between two variables to see if it substantially exceeds the Pearson r for the same variable pair, as an indicator of a second-order relationship, was employed.

Results

Questionnaire data from the study were evaluated using one-way ANOVA and are presented in Table 1 below. Reported here are comparisons of four occupational groupings, all associated with the implementation of the advanced manufacturing systems. These four groups are: 1) managers and engineers, 2) foremen or first-line supervisors associated with the project, 3) operators, and 4) skilled trades maintenance, set-up, or technicians. Fifty-two of the 57 completed questionnaires included occupational job titles needed for these analyses.

Table 1: Occupational Group Comparisons in Deployment of New Process Technology

# Variable, Subscales (Cronbach Alpha)	Group Means [1] Mgrs, Engrs ($n = 16$)	[2] Foremen ($n = 9$)	[3] Operators ($n = 14$)	[4] Skilled Trades ($n = 13$)	ANOVA F	Significant Multiple Comparisons•
Job Satisfaction A (.86)	87.9	84.3	78.4	79.7	2.9*	[1] > [3]
Skill Variety (.59)	13.3	13.1	12.1	12.4	1.0 (n. s.)	
Autonomy (.62)	13.0	12.0	11.4	12.8	2.1 (n. s.)	
Job Satisfaction B (.85)	50.6	44.2	43.4	43.0	4.7**	[1] > [4]; [1] > [3]
Deal with others (.60)	13.0	12.1	11.1	11.5	5.9**	[1] > [3]; [1] > [4]
Role Stress (.90)	50.2	47.7	47.1	39.6	3.6**	[1] > [4]
Ambiguity (.82)	20.2	19.6	18.8	17.6	1.5 (n. s.)	[1] > [4]; [3] > [4]
Conflict (.88)	30.0	28.1	28.3	22.0	4.3**	
Work Climate (.92)	43.4	38.8	37.3	30.9	5.2**	[1] > [4]
Work Group Technology (.90)	32.9	24.6	25.5	26.3	3.8*	[1] > [2]; [1] > [3]

Note: Total cases = 52. Five of the original 57 cases were unusable because there was no job category information.

** $p < .01$

* $p < .05$

• Scheffé Ranges, $p = .10$

Job Satisfaction scales from *Hackman* and *Oldham* (1980); Role Stress from *Rizzo* et al. (1970); Work Climate adapted from *Taylor* and *Bowers* (1969); Work Group Technology from *Withey* et al. (1983).

The four groups were compared on the scales discussed above with Cronbach alpha reliabilities reported in Table 1. Subscales for the *Hackman* and *Oldham* (1980) job diagnostic survey included skill variety, autonomy, and dealing with others, and for role stress, the two dimensions of role ambiguity and role conflict are presented (cf. *Rizzo, House*, and *Lirtzman* 1970: 158).

The groups differ in at least one multiple comparison for all four measures. Predictable patterns were found for job satisfaction. Jobs at higher levels in the hierarchy, managers and engineers, are associated with greater satisfaction than at lower levels, here operators or skilled tradesmen. The same pattern is true for work climate, with the exception that operators and foreman tend to be more alike than skilled tradesmen.

Hypothesis 1 predicts that supervisors will have routine jobs after modernization. Supervisors differ significantly from managers or engineers as do operators, with $F = 3.8$ ($p < .02$). The first-line supervisors report the most routine jobs, which indicates that they have the fewest exceptions and most analyzable tasks, just after new production technology is installed. Hypothesis 1 is supported by these results.

Contrary to Hypothesis 2, supervisors did not have the lowest autonomy, and the four occupational groups did not differ significantly on this subscale ($F =$ 2.1, n.s.). However, skilled tradesmen and managers or engineers were found to have higher autonomy scores, with only operators lagging behind supervisors in this dimension. Findings are generally in the predicted direction, albeit non-significant. In addition, the four groups did not differ significantly on skill variety ($F = 1.0$, n.s.). Hypothesis 2 was not sustained by these results.

Hypothesis 3 predicts that in new system deployment, supervisors are likely to depend more on human relations skills. This hypothesis is moderately supported by the results in Table 1. Managers or engineers were significantly higher on this task dimension than operators and skilled tradesmen ($F = 5.9$, $p < .01$), but not significantly different from first-line supervisors. This suggests that for dealing with others, supervisors are very similar to their next reporting level in manufacturing plants deploying new production technology.

T he plot of workgroup technology and role stress scores for all 57 complete respondents is presented in Figure 1. The zero-order correlation was non-significant with $r = .08$, as expected. A second order regression test was run with the stepwise option entering workgroup technology scores first and then workgroup technology scores squared second, resulting in a significant improvement in explained variance ($\Delta R^2 = .08$) with $F = 4.7$ (df = 1,55, $p < .05$). As can be seen from Table 1, moderate levels of routineness (exceptions and analyzability) result in the lowest levels of overall role stress, while extreme levels of workgroup technology, either all routine or no routine on the new job, result in the highest levels of role stress. Hypothesis 4 is supported by these results, because it predicts this curvilinear relationship.

Returning again to Table 1 for an evaluation of Hypothesis 5, which predicts

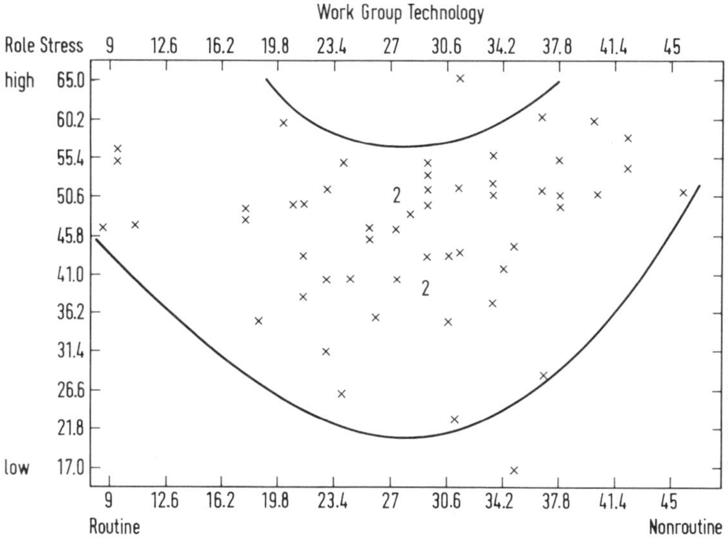

Figure 1: Role Stress vs. Work Group Technology (n=57)

stress will vary by occupation, it was found that there were significant differences in the predicted direction between these four occupational groups on role stress ($F = 3.6$, $p < .01$). As expected, managers or engineers had the highest role stress scores, most likely because they also had the highest workgroup technology scores. Next came first-line supervisors and operators. Lowest were skilled tradesmen, who also had the moderate levels of routineness and analyzability. The only significant multiple comparison was the difference between managers or engineers (average score = 50.2) and skilled tradespeople (average score = 39.6). First-line supervisors (average score = 47.7) and operators (average score = 47.1) did not differ significantly and were just slightly above the grand mean of 45.7 on the role stress scale.

Although no predictions were made concerning the decomposed sections of the role stress scale, role ambiguity and role stress, some interesting results were obtained on these subscales (Table 1). First, there were no significant differences found between these occupational groups on role ambiguity ($F = 1.5$, n.s.). On the other hand, two multiple comparisons were statistically significant for the role conflict subscale. The F-test was conducted first, followed by the multiple comparisons (one-way ANOVA with $F = 4.3$, $p < .01$, cf. Table 1). Both the manager/engineer group and the operator group were found to have significantly greater role conflict than the skilled tradespeople group. First-line supervisors were most like operators in their scores (mean = 28.1 and 28.3, respectively), but did not differ significantly from any other group. The role conflict grand mean was 27.2. These results confirm Hypothesis 5.

Discussion

Four of the five hypotheses concerning the role of the first-line supervisor were generally supported by preliminary questionnaire results from domestic plants currently deploying advanced manufacturing technology. Occupants of first-line positions report the most routine and analyzable jobs and are most like managers or engineers in higher scores for dealing with others, task content, and are above average on role stress. Apparently, first-line supervisors have elevated stress levels for reasons quite different from managers and engineers. Supervisor's jobs are routine while manager's jobs are quite the opposite. Contrary to predictions, the first-line was not found to differ significantly from the other three occupational groups on autonomy in task content, although results are generally in the predicted direction: supervisors tend to have less autonomy.

The emergent picture in these data is reminiscent of the overall evaluation of the supervisor's role of an earlier generation – one that is similar in some respects to levels above and below the position in the hierarchy and rarely an outlier on measurement scales. However, during the early stages of deployment of new production technology, the details of the patterns have become so extreme in one area, i. e., routineness of duties, that careful re-examination of the supervisor's role is now warranted.

Latack et al. (1985) found, for a sample of supervisors, that especially autonomy and to a lesser extent dealing with others and social support were directly and significantly related to choice of the effective coping behaviors of positive thinking and direct action. They also found skill variety to be inversely and significantly correlated with the ineffective coping behavior of avoidance or resignation, while role ambiguity significantly promoted this ineffective choice.

Here it was found for a sample of occupations all intimately involved in the deployment of advanced manufacturing technology that there was no significant difference between groups according to autonomy, and managers or engineers had significantly higher task content on dealing with others than operators and skilled tradespeople. No statistically significant results were found for skill variety or role ambiguity, and no data were available on social support.

From the perspective of previous findings and on the basis of statistical significance alone, it would not appear that supervisors arc particularly vulnerable in dealing with the stresses of new technology deployment. However, one additional finding of the current study casts doubt on that overall conclusion. This significant finding was the extreme position of supervisors on the work-group technology scale. Supervisors were found to have the most routine, analyzable jobs with the fewest exceptions after advanced manufacturing technology was deployed. The extremes in routineness and analyzability were significantly associated with high role stress, which suggests that supervisors, without any particular advantage in dealing with this stress, are vulnerable. In addition, the issue of job security in the face of rapidly changing technology is not adequately captured by the task dimension measures used here and in other research relating stress, coping, and new technology. Many supervisors are simply not around anymore to report their reactions to change.

This important last finding may explain, in part, why the first-line position is being eliminated in some plants undergoing modernization. By giving some of the duties of the old first-line supervisor to the area manager, who presumably has other reasons to choose positive coping behaviors like more opportunities for dealing with others and giving other parts of the former first-line job to operators, a crucial problem in deploying new production technology may be resolved. The problem is that of coping with the stress of learning new skills and potential job insecurity forced by a new technology. In some cases, it may be that the first-line job is not eliminated but expanded with more responsibility and area command. The net effect, however, is really the same.

Implications for Researchers

A number of important questions remain unanswered by these data and analyses. First, we know precious little about the role first-liners play in planning modernization. Case history and anecdotal reports suggest the supervisor can have a considerable informal influence on the important operator and perhaps maintenance personnel selection decisions. The entire issue of first-line participation in the pre-adoption and transition stages of new system deployment needs investigation. It is likely that required best behaviors change, as plants go from startup to steady-state.

A second area that requires exploration and evaluation is the factors that influence the restructuring of the first-line position for modernization. How is it that some plants and not others choose to expand the operator's role with some first-line duties and expand the first-line to an area manager in others? To what extent does a shift in the internal labor market of a plant after modernization influence or moderate this restructuring? These are just some of the questions raised by this study.

Implications for Managers

Although this study should be considered a start rather than the final word to inform the complex task of managing manufacturing firms in technological transition, there are some significant trends worth taking into account when formulating a deployment strategy for the factory of the future. These trends make general recommendations in three categories: 1) the selection and development of personnel, 2) using cross-functional teams for deployment of advanced manufacturing technology, and 3) adoption of new organizational structures for modernization.

Selection and Development of Personnel

The first-line supervisor's job is likely to fall into one of two categories in modern plants. The first type of supervisor's job is likely to be the modern equivalent of the general foreman or area supervisor supported by integrating computer technology. That is, the first-line supervisor is likely to become more like a middle manager in a production organization. The second type of first-line supervisor is that of the manufacturing engineer. The role is heavily dependent on deep technical knowledge of new production systems, often requiring less interaction with line and support personnel than before. This engineer is likely to stay in the job of supervision a limited amount of time, say two to five years, and might be on the promotion list to general management of a competitive, world class manufacturing organization. As software development and maintenance procedures become more standardized and documented, this type of transition management becomes more feasible. Human relations skills are not eliminated, but have perhaps been overemphasized as a success factor in earlier research and cases.

Regardless of plant or type of supervisor, the role occupant will have to be better educated and more experienced. Some present-day supervisors will probably be retrained to fit the general foreman model in a modern plant, but they cannot get this training in the typical supervisory course being offered today. These supervisors need skills in planning and controlling computer-driven technologies, and they need to communicate with technical and functional specialists within and outside of the plant. They will have intimate knowledge of production control, inventory management, technology planning, and a sound grounding in manufacturing strategy. Exposure to functional strategic planning such as marketing and finance, as well as business unit planning, will be necessary.

The manufacturing engineer that makes the transition to line management will have to customize the best of project management style and savvy to line management technique. It is likely that this type of incumbent will rely on and, therefore, need to cultivate more support relationships. With the coming of the engineering generalist, however, there will also be the option available to this person of rotating into other engineering positions in the firm as they open.

New supervisors require new, specialized development courses that are intense in their treatment of manufacturing strategy in the corporate plan. Many universities are beginning to offer such intensive courses, but we have only begun to see the evolution of such specialized educational opportunities. More importantly, the recognition that supervision in modern plants needs to be developed among first-line supervisors as well as their support staff is becoming more widely accepted.

Supervisors on Cross-Functional Deployment Teams

The new first-line supervisor will take his or her place on the cross-functional deployment team in modern plants. This will avoid, in part, the problem that the participative approach to organizational change has suffered in the past: the person actually responsible for supervision of the area where new technology was installed was not included in the planning for the project. Further, this team or committee is often too large and does not get membership on a continuing basis from functions that are key and can commit authority to decisions. Therefore, supervisors should be part of the core deployment team and take an active role in directing the specification effort for all aspects of the program.

New Organizational Structures

Computer-based production technologies tend to integrate functions and break down organizational boundaries. They also have the potential to exacerbate problems that have gone unsolved for many years. Supervisors need support. Organizational structures and procedures that formally link the supervisor with support functions should be considered. Joint quality organizations are becoming common, both locally certified and corporately audited. Rewards should continue to be based on outcomes, but peer support should be encouraged for stress management and development.

Supervisors can now aspire to higher level positions in organizations, especially those that have larger plants. But since the trend is toward smaller, more manageable units and distributed processing computer control which allows decentralized control of manufacturing, the new supervisor will also be required to take on more responsibility, with less autonomy than before. The challenge of the position remains high. It is a difficult job, with variable rewards. If the position is placed correctly in the new plant hierarchy, it can contribute significantly to organizational effectiveness. If the role is forgotten and its occupants ignored, firms will fail to capitalize on the potential benefits of adopting new manufacturing technology.

References

Dill, W. R. (1965). Business organizations, in: *J. G. March* (Ed.), *Handbook of Organizations,* pp. 1071–1114, Chicago: Rand McNally & Company.

Ettlie, J. E. (1985, August): Organizational adaptations for radical process innovations, Paper presented at the 45th Annual Meeting of the Academy of Management, San Diego, CA.

Ettlie, J. E., and *Rubenstein, A. H.* (1980): Social learning theory and the implementation of production innovation, *Decision Sciences,* 11, 4: 648–668.

Gerwin, D., and *Blumberg, M.* (1984, November): Workers' perceptions of their jobs on flexible manufacturing systems, in: *Proceedings,* pp. 175–178, Synergy '84, Chicago, IL.

Hackman, J. R., and *Oldham, G. R.* (1980): *Work Redesign,* Reading, MA: Addison-Wesley.

Haddad, A., *House, J. S.*, and *Jackman, M.* (1985): The impact of technology on industrial work and workers, University of Michigan, working paper.

Hill, K. D., and *Kerr, S.* (1984): The impact of computer integrated manufacturing systems on the first-line supervisor, *Journal of Organizational Behavior Management,* 6, (Fall-Winter): 81–97.

House, R. J., *Schuler, R. S.*, and *Levanoni, E.* (1983): Role conflict and ambiguity scales: Reality of artifacts, *Journal of Applied Psychology,* 2: 334–337.

Iron Age (1986): Plants update, so do foremen, May 16: 47–51.

Latack, J. C., *Joseph, H. J.*, and *Aldag, R. J.* (1985): Job stress: Determinants and consequences of coping behaviors, University of Wisconsin at Madison, working paper.

Lawler, E. E., III. (1978): The new plant revolution, *Organization Dynamics,* Winter, 6 (3): 3–12.

McGrath, J. E. (1970): A conceptual formulation for research on stress, in: *J. McGrath* (Ed.), *Social and Psychological Factors in Stress,* New York: Holt, Rinehart and Winston.

Mann, F. C., and *Hoffman, R. L.* (1956): Individual and organizational correlates of automation, *Journal of Social Issues,* 2: 7–17.

Namboodiri, N. K., *Carter, L. F.*, and *Blalock, J. R.* (1975): *Applied Multivariate Analysis and Experimental Designs,* New York: McGraw-Hill.

Noble, D. F. (1986): *Forces of production,* Oxford: Oxford University Press.

Rizzo, J. R., *House, R. J.*, and *Lirtzman, S. I.* (1970): Role conflict and ambiguity in complex organizations, *Administrative Science Quarterly,* 15: 150–163.

Rosen, M. I., *Aldag, R. J.*, and *Joseph, H. J.* (1985, November): Workplace coping: Development of a structural equation model, in: American Institute of Decision Sciences, *Proceedings,* pp. 616–618, Las Vegas.

Taylor, J. C., and *Bowers, D. G.* (1969): *Survey of Organizations,* CRUSK, Institute for Social Research, The University of Michigan.

Walton, R. E., and *Schlesinger, L. A.* (1979): Do supervisors thrive in participative work systems? *Organizational Dynamics,* (Winter) 7 (3): 25–38.

Withey, M., *Daft, R. L.*, and *Cooper, W. H.* (1983): Measures of Perrow's work unit technology: An empirical assessment and a new scale, *Academy of Management Journal,* 1: 45–63.

Woodward, J. (1965): *Industrial Organization: Theory and Practice,* London: Oxford Press.

Conclusion

A Brief Summary of Volume I

Urs E. Gattiker

Summaries are intended to restate and integrate what has been said earlier. Is it possible or desirable to do that here? The material is complex, the parts lend themselves to independent examination and evaluation, and the potential forms of integration are numerous. Furthermore, the research described in this volume is better viewed as an unfinished venture, and the chapters in this book as progress reports rather than final works. Still, it appears useful to summarize some of these results and identify their practical implications as well as several strategic issues which may guide future research endeavors.

Part I

The authors of Chapter 1 demonstrated that a "liberated" work environment for employees substantially increases a firm's innovation potential. Yet, in order to test this "liberating" form of an organization further, it is necessary to do not only more qualitative but also quantitative studies.

In Chapter 2, it was suggested that joint international research and development (R&D) ventures can substantially reduce a company's R&D costs. However, there are many obstacles in this course of action, such as differences in management style and cultural values between the firms or countries involved. To ensure success, such ventures require extensive negotiations and contract work which can be time-consuming and expensive. The example of cooperation between American and Japanese companies in the semiconductor industry shows that this kind of joint venture has not been as beneficial to the United States as was initially hoped. First, a large amount of intra-cultural as well as cross-cultural research is needed to determine those conditions which actually facilitate the success of such ventures (e.g., *Bhagat* and *McQuaid* 1982).

It was shown in Chapter 3 that innovation in advanced defense technology requires not only competitive pricing but also special assurance strategies, such as technical and delivery records, to close a contract. The authors stated that further research is needed to test their marketing strategy model. For instance, an analysis of industrialization and innovation categories might well suggest that certain strategies are particularly successful in certain situations.

Perhaps the most intriguing discovery for managers was discussed in Chapter 4. The authors suggested that industrial innovation does not necessarily lead to increased profits. Although several other researchers have previously argued along these lines (e.g., *Gold* 1983), this study is unique, because is presents a cross-organizational approach in analyzing some of the largest firms in the United States. It also points out that employment at innovative companies may not necessarily be more secure than with non-innovators. These claims should be tested outside North America as well, to permit greater generalizations.

Research was presented in Chapter 5 which indicated that Canadian low tech firms can and do create as many new jobs as high tech enterprises. Such studies ought to be replicated in other countries to see if the implications of these results hold for most industrialized nations. Incentive packages for high tech companies are politically popular, but it remains to be seen whether such subsidies can, in fact, promote the desired creation of jobs. Low tech industries, on the other hand, can provide employment opportunities in many areas where high tech firms rarely become involved, such as the lumber industry in the American Northwest and Canada. Indeed, vertical integration in the lumber industry may add many jobs to rural areas, thus helping to sustain local economies.

Clearly, organizations must ensure that their employees have the necessary habits, skills and styles from which they can construct the appropriate "strategies of action" to manage innovation effectively (cf. *Swidler* 1986). However, since individuals differ in their characteristics, backgrounds, and learning abilities, diverse training programs are required so that people can learn successfully. Continuous training opportunities for a firm's human resources may well provide one crucial link between successful innovation, R&D, and effective use of new technologies in organizations around the world.

Part II

The papers in this section have much to say to managers, employees, and trade unions. The data presented also raises several issues which should be addressed in future research. Employee attitudes toward technologies comprise an important field of inquiry which needs to be studied closely. There is research indicating that negative attitudes can lead to increased levels of stress (cf. *Kohn* et al. 1983). Work included in this volume suggests that negative attitudes also reduce the desire to utilize technology most effectively. Furthermore, other researchers claim that technology resistance reduces knowledge, while inadequate skills reduce user satisfaction. Thus, the effective application of new technology could be in serious jeopardy (e.g., *Bikson* and *Gutek* 1983).

According to the authors of Chapter 6, managers are using their electronic workstations mainly to increase the efficiency of peripheral activities. Travel, for example, is being substituted by teleconferences in some instances. Yet, the real benefits of such technologies are reaped when they are successfully integrated into core work activites.

Chapter 7 demonstrates that human resource specialists perceive the effect of human resource information systems positively, but that other end-users disagree. Such discrepancies may well be the result of communication problems which must be addressed before an information system can be put to optimum use.

This notion is also supported by the research findings in Chapter 8, which claim that different groups of computer users manifest differing attitudes toward computer-based technology. For instance, women do not feel as positively about computer-mediated work as men do. While it may not be possible to accommodate all needs simultaneously, strategies must be developed and adjusted from time to time to ensure the greatest amount of end-user satisfaction for the largest number of employees.

It is suggested in Chapter 9 that first-line supervisors need assistance in improving their interpersonal skills. Technological innovation requires increased interfacing with subordinates and others in the organization, which in turn necessitates better communication skills. The issue of "man and machine" is usually complex when new technologies are introduced, and managers should take great care to pay as much attention to the personnel needs as they tend to give to the technical aspects.

The effects of technology on employees and their work are vast and not yet well known. Research in this section showed that some employees have negative attitudes toward computer technology. At the same time, employees must change their skill components to keep up with technological development. As a result of these studies, firms should invest in further training of their human resources. Perhaps technology is only being used peripherally because employees lack the necessary application skills to use it in their core job activities. Research by *Bikson* and *Gutek* (1983) confirms his notion; employees in their sample felt inadequately trained to use their technology effectively in their daily work.

Conclusion

Obviously, there are some serious challenges ahead. Future technological innovation may require further adaptation by employees. However, little research has been done to determine what type of changes could occur and, most interestingly, how companies, employees, and their unions can manage technological innovation effectively. Since workers may not always feel satisfied with their technology, the overall quality of work life could deteriorate, which would be highly undesirable. More data is needed in this area to provide the information necessary to make strategic changes which will prevent technology from lowering the quality of work life.

Additional technological innovation will require that individuals re-enter the education process to learn new skills. Such skills are not only necessary for using the technology, but to accommodate structural work changes as well. Further investigation is needed to determine the type of training which will facilitate the acquisition of the necessary job and technology application skills.

As a field, technological innovation and human resources is in its infancy. A great deal more work needs to be done before this area can be more clearly defined. We hope that this volume has provided the reader with a sense of the wide range of territory that is represented in the field, not only at the current "frontier" but also trends for the future. The work which will be accomplished over the next few years may prove crucial both in terms of long-term direction and the nature of technological innovation.

References

Bhagat, R. S., and *McQuaid, S. J.* (1982): Role of subjective culture in organizations: A review and directions for future research, *Journal of Applied Psychology Monograph,* 67: 653–685.

Bikson, T. K., and *Gutek, B. A.* (1983): *Training in automated offices: An empirical study of design and methods,* (Report No. WD-1904-RC), Santa Monica, CA: The Rand Corporation.

Gold, B. (1983): Strengthening managerial approaches to improving technological capabilities, *Strategic Management Journal,* 4: 209–220.

Kohn, M. L., Schooler, C., Miller, J., Miller, K. A. Schoenbach, C., and *Schoenberg, R.* (1983): *Work and personality: An inquiry into the impact of social stratification,* Norwood, NJ: Ablex.

Swidler, A. (1986): Culture in action: Symbols and strategies, *American Sociological Review,* 51: 273–286.

Where Do We Go From Here? Directions for Future Research and Managers

Urs E. Gattiker

Various ideas and approaches for achieving a better integration of technological innovation efforts and strategic human resource management are being discussed today. There are, however, still some critical gaps in our knowledge of this vast field. The following sections will outline in more details where further work is needed to advance our current understanding.

Level of analysis. Technological innovation, and especially R&D efforts, can be analyzed at a number of levels, including those of individual members, groups, organizations, networks of organizations, and even larger communities. Much of the research in this area can be grouped along a micro/macro dimension. While some deal specifically with the end-user, other studies analyze the effect of R&D efforts and corporate strategies of organizations on profit and sales.

Although his micro/macro distinction is useful for incorporating a number of issues into the technology innovation and human resources domain, it also has its shortcomings. For one, there is a tendency to treat micro/macro concerns as separate problems instead of interrelated issues. Yet, to increase our knowledge about the technological innovation process and its relationship with human resources (e.g., beliefs, attitudes, type of use, organizational culture), it is necessary to examine different levels of analysis in one study. Thus, R&D efforts and their effects on innovation as well as quality of work life for a company's human resources should be studied by combining a micro and macro approach.

Organizational adaptation to technological innovation. Technology adaptation may be most easily understood by contrasting internal and external contexts. *Pfeffer* (1982: 162) argued that considerable importance should be attached not only to external organizational facts – size, technology, and environment – but also to internal processes. Specifically, more attention needs to be given to the development of a typology which, most of all, would allow measuring those internal processes which affect technological change.

The research presented in this book is a big step in the right direction because the focus is on the effects of size, technology, and environment upon technological innovation as well as internal factors, such as a firm's human resources. Further work is needed which creates concepts that can be applied across organizations and even countries. In addition, standardized measures should be developed which facilitate comparisons of different studies.

Developing standardized, comparable instruments. It seems clear that if research is to be cumulative, a great deal more standardization is necessary in all aspects of measurement. There is little agreement on the best ways to measure technological innovation and its outcomes (e.g., effectiveness, return on investment, quality of work life). This absence of standardization can be attributed to a lack of congruence between the disciplines of management, sociology, and psychology as to what should be measured and how to measure it. Clearly, some consensus is needed on how to measure basic variables of technological innovation, such as R&D investments and effectiveness, corporate strategy, work skills, and end-user needs. As it is, each time technological adaptation is assessed, a re-invention of the wheel appears necessary for the purpose of collecting data. Standardized instruments would greatly facilitate cross-site comparisons and allow for generalizations and new insights.

Theory building, theory testing. A data base developed with standardized instruments would provide a rich foundation for the development of theories about managing technological innovation in organizations. First, however, it is necessary to have measures which include several levels of analysis (macro and micro), are broad in scope, and include different data gathering techniques (qualitative and quantitative). Additionally, both objective and phenomenological characteristics need to be included, and longitudinal data should be used.

A serious attempt should be made to prune away the intellectual dead wood; we have become far too committed to our own perspectives. As a result, disconfirming evidence often becomes accommodated by introducing new intervening variables and contingencies instead of looking for viable alternatives that may explain the phenomena in a fairly simple and accurate fashion. The contributions in this book include both qualitative and quantitative studies. Some chapters are longitudinal, but most use a cross-sectional approach. Future research should continue along these lines with courage and conviction.

Corporate strategy and R&D efforts. A lot of attention has been paid to the effects of corporate strategies on profitability, R&D efforts, and innovation. Yet, as *Pfeffer* (1981) already pointed out while reviewing some of the organizational assessment literature, we must learn from those instances in which prophecy fails and effects are not found. Researchers should look for a small set of powerful concepts that are relatively simple in their application for the measurement of R&D efforts and innovation. Those efforts and their limited effect on profitability may require companies to change their strategies. Similarly, governments might have to reassess their support of companies which appeared full of promise in the area of job creation but then could not deliver. Some high tech firms are a case in point.

Job skills and technological innovation. The debate over the effects of technology on skills, work, and employees, as well as the changing nature of the labor process, have shadowed industrial development for centuries. The issues correspond to the notions of upgrading, downgrading, mixed effects, or relative net change in the skill levels of work over time (*Spenner* 1983). There is no verdict yet on these arguments, and the central question remains: which are the skills that may indeed help in coping with technological innovation?

Any discussion of skills must also involve the range and depth issue. There is a linkage between skill requirements for certain positions and specialization of employees. For instance, a pediatric nurse will have a great depth of skills in this area but may have a limited range of skills in emergency care. It would be difficult for the nurse to transfer from one area of expertise to the other without adequate supplemental training.

Training of end-users. Training must be designed in such a way that individuals can easily relate newly acquired knowledge to appropriate job situations. They should be capable of using a new technology innovatively in solving various tasks, therefore, they should be encouraged to master relevant challenges during the period of training. The learning goal must focus on either utilizing or activating the employee's own ability to overcome obstacles (*Dweck* 1986). Innovation thus requires continuous training of the workforce to ensure effective use of new technology.

New approaches for technology training have already presented themselves. Video disks are being used to train and educate workers. Computer-aided learning (CAL) has also become popular in combination with the more traditional avenues, such as studying a manual or using a laboratory setting for learning exercises. The issue here is not to determine which of the methods is best, but to facilitate the learning process for employees with different levels of intellectual ability. Recent research found that university students with low academic ability gained most from CAL when trying to acquire computer application skills (e.g., *Gattiker* 1987, *Gattiker* and *Paulson* 1987). Job-related adult education requires the development of innovative teaching methods which can be used in a variety of organizations to train individuals for newly introduced advanced technology. Training time is precious, and positive results become mandatory to justify the expenditure of both hours and financial resources.

Conclusion. Technological change and human resources is not a totally new field; people have been doing relevant research for decades. Clearly, a number of advantages can be expected from an integration and accumulation of research. The benefits that have already been reviewed are highly desirable. *Technological innovation and human resources* should be the name of a whole new area of research endeavours, as initiated or perpetuated by this book. In this context, technological innovation should go beyond computers and robotics to encompass all new technology, such as genetics, lasers, space technology, etc. When done well, further work in the area of technological innovation and human resources can produce valuable information of a kind that has never been gathered before. There is also the potential to combine this information in new and exciting ways, and to reach groups of technology end-users that have not previously had access to such knowledge.

Several points can be made about how and where technological change in an organization can be most successful in terms of human resources:

1) In order to recruit individuals who closely fit the organization and its culture, it is necessary to improve personnel selection and testing. Trial and error is much too costly, as turnover becomes ever more expensive for companies. An extensive human resource accounting system ought to be in place to allow management to quantify the cost of errors in recruiting. Thus the selection instruments must be validated before they can be used effectively.

2) Both first-line managers and support personnel must assume some responsibility and participate in the effective management of technological change. To achieve this end, the employee needs periodic training which should allow him/her an opportunity to acquire the necessary skills for using new technology innovatively in daily work. The training should be away from one's workplace and occur over several weeks (cf. *Gattiker*, in press). Compact training over a few days or a weekend may lead to overload, thereby reducing the effectiveness of the program.

The substantial investment in training of a firm's employees makes their long-term employment within the company highly desirable. At the same time, a company will allocate resources for training more readily if it can expect to benefit from a loyal and committed workforce. Participative career planning and management by both the employee and the organization thus becomes imperative.

3) To stimulate innovation, an organization's culture should encourage independent thinking on the part of each employee and participation in the decision-making process at the lowest level. Such efforts may take a few years before they are deemed successful and the results justify this kind of approach. When risks are taken, mistakes will happen which can be costly. It would be unfortunate, however, if management would then revert to the status quo and become risk-avoiders.

4) End-user needs should be taken into careful consideration when leaping forward in the information systems area. Information systems specialists and end-users must communicate effectively so that the information flow is conducive to the most productive and profitable utilization of such systems.

5) Companies need to assess the beliefs of their employees about new technologies before their introduction as well as during their utilization. Helping individuals to better understand what is happening to their working environment will prove beneficial to the employer over time. Necessary changes, as identified in management-worker dialogue, can then be implemented successfully to improve the effective use of any new technology.

6) It is in the best interest of an organization to look for joint R&D ventures which might lead to synergy effects, thus reducing costs and making programs more effective. A cultural match between employees of the firms involved must be accomplished to develop innovative products.

7) Effective innovation requires a sophisticated marketing approach which not only takes into careful consideration the type of technology to be offered but especially the end-user (government vs. private business) to allow rapid diffusion of the innovation. The greatest invention is of little use to the business community if it is not readily available for implementation at the workplace.

8) Governments need to seriously re-assess their support of high tech industries. Subsidies, tax write-offs, and other incentives may not lead to the desired creation of jobs. Unfortunately, a reallocation of public funds may require admission of failure which could be harmful to political careers. It is often easier to throw good money after bad in maintaining popular ventures, even if their worth to the economy has become questionable.

Effective management of technological innovation is possible. If we continue to improve our understanding of this important concept, new methods can be developed to increase our effectiveness in R&D and strategic human resource management. We should refrain, however, from jumping onto band wagons and be careful not to overlook what we have already learned, simply because it is called by a new name. If the current interest in technological innovation and human resources is taken seriously, it should assist us into the next century without any threatening declines. An integrated approach to managing organizations is only possible when the workforce is placed at center stage, and technological innovation becomes a true partner rather than an overriding concern.

References

Dweck, C. S. (1986): Motivational processes affecting learning, *American Psychologist,* 41: 1040–1048.

Gattiker, U. E. (1987, May): Identifying efficient training methods for teaching micro-computer skills to adults: Academic ability and learning performance, Paper presented at the International Conference on Computer-Aided Learning in Post-Secondary Institutions, Calgary, Canada.

Gattiker, U. E. (in press): Technological adaptation: A typology for strategic human resource management, *Behaviour and Information Technology.*

Gattiker, U. E., and *Paulson, D.* (1987): Testing for effective teaching methods. Achieving computer literacy for end-users, *INFOR Canadian Journal of Information Systems and Operations Research,* 25: 256–272.

Pfeffer, J. (1981): Four laws of organizational research, in: *A. H. Van de Ven* and *W. F. Joyce* (Eds.), *Perspectives on organization design and behavior,* New York: John Wiley and Sons, 409–418.

Pfeffer, J. (1982): *Organizations and organization theory,* Boston: Pitman.

Spenner, K. I. (1983): Deciphering Prometheus: Temporal change in the skill level of work, *American Sociological Review,* 48: 824–837.

About the Contributors

Frank L. Edwards is Advanced Programs Manager, Hughes Aircraft Company. He received his doctoral degree at Claremont Graduate School in the Executive Management Program in 1984.

Mailing address: Advanced Programs Manager, Hughes Aircraft Company, Fullerton, CA 92634, USA.

John E. Ettlie is the Director of the Office of Manufacturing Research in the School of Business Administration at the University of Michigan. He earned his Ph.D. in Industrial Engineering at Northwestern University in 1975. He has held several research, consulting and university positions and has published over 40 articles on the management of technological change. His current research and administrative activitives include the direction of a four-year study on the deployment strategies for advanced manufacturing technologies in domestic plants, funded by the National Science Foundation.

Mailing address: Office of Manufacturing Research, School of Business Administration, University of Michigan, Ann Arbor, MI 48106, USA.

Steven W. Floyd is an Assistant Professor of Management at the University of Massachusetts at Amherst. He received his Ph.D. from the University of Colorado in 1986. His research focuses on managerial use of information technology and the strategic management of information technology. He has made presentations before the Academy of Management and the Decision Sciences Institute. Before his recent transition to academe, Dr. Floyd was President of a medium-sized manufacturing firm.

Mailing address: School of Management, University of Massachusetts at Amherst, Amherst, MA 01003, USA.

Urs E. Gattiker is Assistant Professor of Organizational Behavior and Information Systems at the School of Management, The University of Lethbridge in Lethbridge, Alberta, Canada, where he also heads the Technology Research Group. He received his Ph.D. in Organization and Management from Claremont Graduate School. His areas of specialization and prior publications include career development, end-user computing, technological and organizational assessment, as well as management consulting.

Mailing address: School of Management, The University of Lethbridge, Lethbridge, Alberta, T1K 3M4, Canada.

B. W. Gibbs received his Ph.D. from the University of Toronto in 1987. He is currently Assistant Professor in the Faculty of Business Administration at Simon Fraser University, Vancouver, British Columbia. Over the past several years, Professor Gibbs has done extensive research on the relationship between managerial roles and organizational effectiveness. He is currently conducting research in the areas of legitimacy and institutionalization in organizations, charismatic leadership, women and power, and is also studying the management of not-for-profit organizations.

Mailing address: Faculty of Business Administration, Simon Fraser University, Vancouver, British Columbia, V5A 1S6, Canada.

Igor Y. Goryunov is a faculty member at the Moscow Institute of Management located in Moscow, Soviet Union. He joined the faculty after receiving his terminal degree from the Institute. He served for a year as a Visiting Scholar in the Department of Management at the University of Texas at Arlington. His research interests focus on the management of research and development to produce effective innovation.

Mailing address: Moscow Institute of Management, Moscow, Soviet Union.

Michael A. Hitt is the T. J. Barlow Professor and Head of the Department of Management at Texas A&M University. He previously served on the faculties of Oklahoma State University and University of Texas at Arlington. He holds a DBA from the University of Colorado. He is the author of six books and over 60 publications in such journals as *Academy of Management Journal, Strategic Management Journal, Journal of Applied Psychology,* and *Decision Sciences.* He serves as a member of the editorial review board for the *Academy of Management Journal.* His research interests focus on effective implementation of strategic decisions, making managerial selection decisions and strategic decisions regarding the management of organizational innovation (R&D).

Mailing address: College of Business Administration, Department of Management, Texas A&M University, College Station, TX 77843-4113, USA.

R. Duane Ireland is the Curtis Hankamer Chairholder of Entrepreneurship and Chairman, Department of Management in the Hankamer School of Business, Baylor University. His teaching and research interests focus on strategic decision making processes in large and small organizations. Among other topics, his research has examined relationships between areas emphasized within a particular firm and the financial performance achieved by that firm and the management of organizational innovation. His publications appear in a range of journals, including the *Academy of Management Executive,* the *Strategic Management Journal, Administrative Science Quarterly, Decision Sciences, Human Relations,* the *Journal of Management Studies,* and *Long-Range Planning.*

Mailing address: Hankamer School of Business, Department of Management, Baylor University, Waco, TX 76798, USA.

K. J. Keen is a Queen Elizabeth II and Ontario Scholar currently completing a Ph.D. in Statistics at the University of Toronto. His doctoral thesis is concerned with the estimation of multivariate correlations in family surveys. Mr. Keen is continuing to investigate topics in the advanced theory of statistics, especially bootstrap methods and multivariate time-series analysis. Based on his concern for the interaction of science, technology, and society, Mr. Keen has been a member of the Canadian Student Pugwash organization for the past six years.

Mailing address: Department of Mathematics, University of Toronto, Toronto, Ontario, M5S 1A2, Canada.

Laurie Larwood is Dean, School of Business, State University of New York at Albany. She is Editor of *Group and Organization Studies, Women and Work,* and *Journal of Management Case Studies.* Her recent books include *Organizational Behavior and Management* and *Women's Career Development.* At the time of writing, she chaired the Academy of Management's R&D/Technology/Innovation Interest Group.

Mailing address: School of Business, State University of New York at Albany, Albany, NY 12222, USA.

R. G. Lucas received a Ph.D. from Cornell University in 1984, and is currently Associate Professor of Organizational Behaviour and Industrial Relations in the Faculty of Administrative Studies at York University, Toronto, Ontario. Professor Lucas spent several years in private and public sector management prior to returning to academia. Since that time, he has published several papers describing his research into organizational politics, decision making, and innovation. Sam the Dog, a black-and-white Husky, has recently joined him in his endeavours.

Mailing address: Faculty of Administrative Studies, York University, Toronto, Ontario, M3J 1P3, Canada.

Jon Olson is a research Scientist at the Battelle Memorial Institute's Human Affairs Research Centers in Seattle, Washington. A graduate of the Sociology program and USC, Dr. Olson's current research interests include Japanese organizations, cooperative ventures, and organizational safety. Dr. Olson also serves as a co-principal investigator in an NSF funded project to study U.S./Japanese cooperative ventures.

Mailing address: Battelle Memorial Institute, 4000 N.E. 41 Street, Seattle, WA 98105, USA.

Richard N. Osborn is Professor of Management and Organizational Sciences at Wayne State University. He has published extensively in the areas of leadership, organization theory, organizational safety, and, most recently, multinational cooperative ventures. Dr. Osborn is co-principal investigator on an NSF funded project to examine U.S./Japanese cooperative ventures.

Mailing address: Department of Management, Wayne State University, Detroit, MI 48202, USA.

Lee Tom Perry is an Associate Professor of Organizational Behavior at Brigham Young University. He previously taught at Pennsylvania State University and Purdue University. While at Penn State, Professor Perry received the Outstanding Faculty Award (1984–85) from the College of Business Administration. He has published numerous articles about industry decline, mergers, and acquisitions, and more recently, radical product innovation. His articles have appeared in journals such as *Sloan Management Review, Organizational Dynamics, Human Resource Management,* and *Personnel.*

Mailing address: Department of Organizational Behavior, 790 TNRB, Brigham Young University, Provo, UT 84602, USA.

Kurt W. Sandholtz was introduced to the study of innovation while working toward a master's degree in organizational behavior at Brigham Young University, where he was involved in much of the field research reported in this essay. His background in English literature and experience as a jazz pianist were crucial in framing the essay's guiding metaphors. Kurt is currently a staff writer for the National Business Employment Weekly, a Dow Jones & Co. publication based in Princeton, New Jersey.

Mailing address: National Business Employment Weekly, P.O. Box 300, Princeton, NJ 08543, USA.

Aubrey Strickstein is a doctoral candidate in the industrial/organizational psychology program at Wayne State University. He has recently conducted research on career development, emphasizing the role of the supervisor. Other academic interests include management of international ventures, corporate values, leadership styles, and mentoring practices in organizations. He has served as a consultant for several large organizations.

Mailing address: Department of Management, Wayne State University, Detroit, MI 48202, USA.

Jo Ann Verdin is a principal partner at Decision Technology Associates, Inc. of Chicago. She specializes in consulting in the areas of computer utilization by managers and professionals, as well as assessing the impact of information systems and computer technology on organizations and individuals. Verdin is currently president of the Midwest Human Resource Planners Group and serves on the research committee of the Human Resource Planning Society. She has a Ph.D. in Management from the University of Houston.

Mailing address: Decision Technology Associates, Inc., 81 E. Elm Street, Chicago, IL 60611, USA.

Author Index

Page numbers marked with an asterisk (*) indicate the location of the full reference, all other page numbers refer to citations within the text.

Abernathy, W. J., 9, 15, 16, 29, 30*, 74, 86, 90*, 91*, 95, 98, 117*
Achrol, R. S., 58, 71*
Ackelsberg, M. R., 76, 91*
Aiken, M., 34, 35, 52*
Al-Bazzaz, S. 79, 90*
Aldag, R. J., 190, 202, 206*
Aldrich, H. E., 35, 52*, 126, 141*
Allen, M. K., 145, 146, 158*
Amico, A. M., 144, 158*
Anderson, J. C., 133, 139*
Andrew, J., 132, 139*
Argyris, C., 47, 52*, 124, 128, 139*
Avard, S., 76, 90*
Azumi, K., 74, 91*

Babin, J. C., 58, 71*
Bagozzi, R. P., 133, 139*
Bailey, R. W., 128, 139*
Bair, J. H., 127, 139*
Ballon, R. J., 41, 52*
Bamforth, K. W., 34, 53*
Bandura, A., 162, 165, 166, 183*
Barbosa, R. R., 77, 80, 90*
Barkin, S. R., 136, 137, 140*
Barley, S. R., 22, 31*
Barnard, C. I., 126, 140*
Bartels, R., 41, 52*
Bass, B., 34, 52*
Baumgartner, J. S., 56, 71*
Baysinger, B. D., 77, 90*
Bearden, W. O., 134, 140*
Bedeian, A. G., 136, 141*
Bell, N. E., 145, 146, 159*, 177, 182, 185*
Bentler, P. M., 136, 140*
Berger, D. E., 163, 164, 165, 166, 167, 168, 179, 181, 184*
Bettis, R. A., 80, 90*
Beveridge, J. M., 59, 71*
Beyer, J. M., 45, 47, 53*
Bhagat, R. S., 209, 212*
Bikson, T. K., 144, 145, 146, 152, 155, 156, 158*, 163, 181, 184*, 210, 211, 212*
Blackler, H. M., 181, 183*
Blake, R. R., 40, 52*
Blalock, H. M., 167, 183*
Blalock, J. R., 198, 206*
Blau, P. M., 125, 140*
Blumberg, M., 195, 196, 206*
Boise, W., 99, 118*
Bonnett, D. G., 136, 140*
Bowers, D. G., 198, 199, 206*
Bowman, E. H., 80, 90*
Boylan, M., 94, 95, 98, 112, 117*
Boyland, M. G., 34, 37, 53*
Bozeman, B., 86, 90*
Branch, B., 80, 90*
Brett, J. M., 133, 140*
Brigham, E. F., 79, 92*
Brousseau, K. R., 165, 178, 183*
Brown, C., 181, 183*
Bryson, J., 37, 52*
Buffa, E. S., 37, 52*
Burns, T., 34, 36, 52*
Business Week, 74, 75, 90*
Buss, M. J., 128, 140*
Buzzell, R. D., 79, 92*

Cameron, K. S., 58, 71*
Campbell, D. T., 130, 140*
Carter, N. M., 125, 140*, 177, 182, 183*
Carter, L. F., 198, 206*
Catto, V., 76, 90*
Cheney, P. H., 125, 140*
Chervany, N. L., 128, 140*
Child, D., 171, 172, 183*
Christensen, H. K., 56, 59, 71*
Cirillo, D. J., 128, 140*

Clark, K. B., 9, 15, 30*, 86, 90*
Clark, M. J., 11, 30*
Clarke, T., 97, 117*
Clausen, J. A., 145, 146, 159*, 177, 182, 185*
Clinard, M. B., 69, 71*
Coe, L., 145, 146, 156, 158*
Cohen, J., 173, 174, 183*
Cohen, P., 173, 174, 183*
Connor, P., 38, 53*
Cook, T. D., 130, 140*
Cooper, R. B., 73, 76, 86, 90*
Cooper, R. G., 26, 30*
Cooper, W. H., 195, 198, 199, 206*
Couger, J. D., 128, 140*
Crookell, H., 95, 117*
Crow, M., 86, 90*

Daft, R. L., 42, 45, 52*, 53*, 195, 198, 199, 206*
Dalton, G. W., 23, 30*
Daly, D. J., 113, 117*
Damanpour, F., 95, 117*
Davidson, M., 76, 90*
Deal, T., 47, 52*
Demby, E. H., 164, 183*
DeMelto, D., 95, 96, 99, 117*
Dewhirst, H. D., 73, 90*
Dickson, D. N., 74, 90*
Dickson, G. W., 128, 136, 137, 140*
Dierkes, M., 182, 183*
Dill, W. R., 188, 189, 193, 205*
Donaldson, G., 76, 90*
Doswell, A., 163, 183*
Downs, G., 94, 95, 118*
Duncan, R., 56, 72*
Durand, D. E., 124, 127, 140*
Dweck, C. S., 215, 217*
Dyer, D., 96, 97, 118*

Economics Encyclopedia, 79, 90*
Edwards, F. L., 58, 62, 71*
Emmelhainz, M. A., 145, 146, 158*
Ennis, D. M., 85, 86, 91*
Ettlie, J. E., 190, 191, 206*
Etzioni, A., 22, 31*
Evan, W., 95, 117*
Eveland, J. D., 34, 37, 53*

Fahey, L., 56, 59, 71*
Falbe, C. M., 125, 140*
Fayol, H., 126, 140*
Feigenbaum, E. A., 60, 71*
Fine, G. H., 74, 90*
Fishbein, M., 163, 183*
Floyd, S. W., 124, 127, 140*
Form, W., 164, 166, 181, 183*
Formbrun, C. J., 46, 52*
Freeman, C., 97, 118*
Freeman, H. E., 162, 184*
Friesen, P. H., 77, 91*
Fuerst, W. L., 125, 140*

Gale, G., 80, 90*
Galitz, W. O., 128, 140*
Gallagher, A., 58, 59, 71*
Gattiker, U. E., 145, 146, 156, 158*, 162–168, 179, 181, 182, 183*, 184*, 215, 216, 217*
Geller, V., 163, 164, 184*
Gerbing, D. W., 133, 139*
Gerwin, D., 195, 196, 206*
Ginzberg, M. J., 125, 128, 140*, 142*
Giuliano, V., 144, 158*
Gold, B., 210, 212*
Goldman, J. E., 9, 10, 31*
Gorbachev, M., 111, 118*
Gordon, G., 58, 59, 71*
Grinyer, P. H., 79, 90*
Grossman, M., 144, 158*
Gutek, B. A., 144–146, 152, 155, 156, 158*, 163, 164, 165, 166, 167, 168, 179, 180, 181, 183, 184*, 210, 211, 212*

Hackman, J. R., 178, 184*, 198, 199, 200, 206*
Haddad, A., 196, 206*
Hage, J., 34, 35, 52*, 74, 91*
Hambrick, D. C., 73, 77, 80, 90*
Hamilton, S., 125, 140*
Hanada, M., 37, 39, 41, 44, 53*
Hannan, R. L., 126, 141*
Harrington, D. R., 80, 90*
Harris, P. R., 144, 158*
Harris, D. L., 144, 158*
Hartwick, J. M., 73, 86, 90*

Hay, K., 95, 97, 99, 118*
Hayes, R. H., 74, 86, 91*
Heany, D. F., 79, 92*
Herker, D., 35, 52*
Hetzner, W. A., 34, 37, 53*
Hill, K. D., 192, 193, 195, 206*
Hiltz, S. R., 127, 140*
Hitt, M. A., 76, 77, 79, 82, 83, 84, 85, 91*
Hladik, K. J., 40, 52*
Hlavacek, J. D., 16, 31*
Hoff, R. D., 144, 158*
Hoffman, R. L., 195, 206*
Hofstede, G., 40, 41, 45, 50, 52*
Holland, W. E., 35, 52*
Horrock, N., 57, 71*
House, J. S., 196, 206*
House, R. J., 188, 198, 199, 200, 206*
Howard, N., 41, 52*
Hrebiniak, L. G., 79, 92*
Hull, F., 74, 91*
Hunt, J., 35, 53*

Iacono, S., 163, 184*
Ireland, R. D., 76, 77, 79, 82, 83, 84, 85, 91*
Iron Age, 188, 206*
Ives, B., 125, 140*
Iyer, E. S., 74, 91*

Jackman, M., 196, 206*
James, L. R., 133, 140*
Jenson, R., 36, 52*
Jewkes, J., 97, 118*
Johnson, E. L., 34, 37, 53*
Jöreskog, K. G., 136, 140*
Joseph, H. J., 190, 202, 206*

Kagan, J., 162, 184
Kahn, R. L., 161, 165, 177, 182, 184*
Kaiser, K., 125, 141*
Kamien, M., 95, 118*
Kantrow, A. M., 86, 90*
Karnani, A., 60, 72*
Keen, P. 127, 141*
Keller, R. T., 35, 52*
Kelly, T. M., 77, 84, 91*
Kennedy, A., 47, 52*
Kerr, S., 192, 193, 195, 206*
Khandwalla, P. N., 79, 91*
Kiesler, S., 163, 164, 184*
Kim, J.-I., 56, 72*
Kimmerly, W. C., 76, 91*
King, B. F., 80, 91*
Kipnis, D., 59, 72*
Kirchner, J., 123, 141*
Klein, K. J., 84, 92*, 95, 98, 119*
Klein, R. E., 162, 184*
Kling, R., 163, 164, 184*
Kohn, M. L., 165, 182, 184*, 210, 212*
Korman, A. K., 162, 166, 184*
Kotler, P., 58, 71*
Kotter, J. P., 126, 139, 141*
Kuhn, T. S., 9, 31*
Kurke, L. B., 126, 141*

Langetieg, T. C., 80, 91*
Lapointe, J. R., 144, 156, 158*
Larwood, L., 59, 71*, 162, 164, 165, 166, 167, 180, 183*, 184*
Latack, J. C., 190, 202, 206*
LaVan, H., 144, 158*
Lawler, E. E., III, 189, 206*
Lawrence, W., 96, 97, 118*
Lawyer, J. Q., 58, 71*
Levanoni, E., 188, 206*
Levine, S. J., 76, 91*
Link, A., 86, 90*
Linn, R. A., 75, 91*
Lirtzman, S. I., 198, 199, 200, 206*
Lloyd, E. J., 161, 184*
Long, S. J., 133, 141*
Lorsch, J. W., 76, 90*
Lubatkin, M., 80, 91*
Lucas, H. C., 123, 125, 128, 137, 142*
Lucas, R., 94, 122, 118*
Lynd, R., 46, 52*

MacMillan, I. C., 55, 57–59, 60–62, 64, 68, 70, 71*, 73, 77, 80, 90*, 91*
Magnus, M., 144, 158*
Maitland, L., 43, 52*
Mankin, D., 144–146, 152, 155, 156, 158*, 163, 181, 184*
Mann, F. C., 195, 206*
Mansfield, E., 85, 91*, 96, 99, 118*

March, J. G., 45, 53*
Marcus, A., 94, 95, 118*
Marcus, L. M., 125, 141*
Mason, R. O., 132, 141*
Mathys, N., 144, 158*
McCaffery, M. L., 77, 91*
McCall, M. W., Jr., 126, 141*
McClenon, P. R., 58, 71*
McCorduck, P., 60, 71*
McDonald, P. R., 59, 71*
McGinnis, M. A., 85, 91*
McGrath, J. E., 195, 206*
McIlveen, J., 95, 118*
McKinley, W., 125, 140*
McMillan, C., 99, 118*
McMullen, K., 95, 96, 99, 117*
McQuaid, S. J., 209, 212*
Meckling, R., 36, 52*
Megaw, E. D., 161, 184*
Mendel, S., 85, 86, 91*
Meyers, M. G., 58, 71*
Mezzrow, M., 11, 14, 31*
Miles, R. E., 10, 31*
Miles, R. H., 58, 71*
Miller, D., 77, 91*
Miller, J., 165, 182, 184*, 210, 212*
Miller, K. A., 165, 182, 184*, 210, 212*
Minasian, J., 99, 118*
Miner, J. B., 126, 139, 141*
Mintzberg, H., 126, 139, 141*
Mirvis, P., 163, 166, 178, 185*
Mitchell, T. R., 46, 53*
Mock, G., 144, 145, 146, 148, 152, 155, 156, 159*
Mohr, L., 94, 95, 118*
Moody, H. G., 127, 141*
Moore, W. L., 9, 10, 15, 31*
Morgall, J., 164, 180, 181, 184*
Morrison, A. M., 126, 141*
Moses, A., 163, 164, 184*
MOSST (Ministry of State for Science and Technology), 98, 112, 113, 118*
Mouton, J. S., 40, 52*
Mulaik, S. S., 133, 140*
Mumford, E., 127, 141*

Namboodiri, N. K., 198, 206*
National Science Foundation, 85, 86, 92*
National Science Board, 85, 91*
Nelligan, T. W., 145, 146, 156, 158*, 164, 165, 184*
Nelson, R., 95, 118*
Nelson, R. R., 77, 84, 92*
New Product Development, 88, 92*
Noble, D. F., 193, 194, 206*
Nogal, G., 144, 158*
Nulty, P., 123, 141*

O'Shea, J., 57, 71*
Ohmae, K., 75, 92*
Oldham, G. R., 178, 184*, 198, 199, 200, 206*
Oliverson, M., 59, 60, 72*
Olson, J., 37, 39, 41, 44, 53*
Olson, J. P., 45, 53*
Olson, M. H., 125, 141*
Osborn, R. N., 35, 37, 38, 39, 41, 44, 53*
Oskamp, S., 183, 184*
Ouchi, W. G., 43, 44, 53*

Pagano, A. M., 144–147, 149, 152, 156, 157, 158*, 159*
Page, R. A., 17, 31*
Palda, K., 111, 113, 118*
Palia, K. A., 79, 91*
Panko, R. R., 163, 184*
Parsons, T., 46, 53*
Paulson, D., 215, 217*
Pava, C. H. P., 127, 141*, 164, 177, 184*
Pavitt, K., 99, 118*
Pazderka, B., 111, 113, 118*
Perrow, C., 35, 53*
Perry, L. T., 17,, 31*
Perry, N. S., 46, 53*
Peters, T. J., 26, 31*, 86, 92*
Pfeffer, J., 45, 46, 53*, 59, 72*, 213, 214, 217*
Podgorecki, A., 179, 184*
Poppel, H. L., 123, 127, 137, 141*
Porter, M. E., 77, 86, 89, 92*
Postner, H., 98, 112, 118*

Quillard, J. A., 144, 145, 146, 148, 152, 155, 156, 159*

Quinn, J. B., 9, 10, 26, 29, 31*

Raelin, J. A., 37, 53*
Ramaprasad, A., 74, 91*
Rappaport, A., 79, 92*
Reilly, F. K., 79, 92*
Research Management, 85, 92*
Rizzo, J. R., 198, 199, 200, 206*
Robertson, T. S., 9, 31*
Rockart, J. F., 144, 145, 146, 148, 152, 155, 156, 159*
Rogers, E. M., 56, 72*
Roitman, T., 34, 37, 53*
Romney, A. K., 162, 184*
Ronan, S., 40, 43, 53*
Rosen, M. I., 190, 206*
Rosenbaum, J. E., 161, 184*
Rowe, L., 99, 118*
Rubenstein, A. H., 190, 206*
Rugman, A., 95, 118*

Salancik, G. R., 45, 46, 53*
Salzman, H., 163, 166, 178, 185*
Sandholtz, K. W., 17, 31*
Sawers, D., 97, 118*
Schein, E. H., 22, 31*, 47, 50, 53*, 181, 185*
Scherer, F. M., 85, 92*, 98, 113, 118*
Schlesinger, L. A., 189, 190, 196, 206*
Schmitt, R. W., 84, 92*
Schmitt, N., 136, 141*
Schneider, J., 34, 37, 53*
Schoeffler, S., 79, 92*
Schoenbach, C., 165, 182, 184*, 210, 212*
Schoenberg, R., 165, 182, 184*, 210, 212*
Schooler, C., 165, 182, 184*, 210, 212*
Schuler, R. S., 188, 206*
Schultz, R. L., 128, 142*
Schwartz, N., 95, 118*
Science Indicators, 74, 92*
Scott, W., 46, 53*
Segal Quince & Partners, 16, 17, 22, 31*
Senn, J. A., 128, 140*
Sharma, S., 134, 140*
Shenkar, O., 40, 43, 53*
Shindler, P. E., 123, 127, 142*
Shneiderman, B., 128, 142*
Smith, T., 162, 184*
Snow, C. C., 10, 31*, 79, 92*
Song, J., 88, 92*
Sonnenfeld, J., 69, 72*
Sörbom, D., 136, 140*
Spenner, K. L., 165, 166, 181, 185*, 215, 217*
Srinivasan,, A., 125, 141*
Stadter, G., 76, 82, 91*
Stalker, G. M., 34, 36, 52*
Starbuck, W. H., 45, 53*
Statman, M., 75, 86, 92*
Staw, B. M., 145, 146, 159*, 177, 182, 185*
Stillerman, R., 97, 118*
Swidler, A., 210, 212*

Tapscott, D., 128, 142*
Taylor, J. C., 198, 199, 206*
Taylor, H. M., 58, 71*
Teel, J. E., 134, 140*
Termoto, Y., 41, 52*
Thompson, V. A., 16, 31*
Thompson, P. H., 23, 30*
Thompson, J. D., 35, 45, 53*, 97, 119*
Thurow, L., 99, 119*
Tornatzky, L. G., 34, 37, 53*, 84, 92*, 95, 98, 119*
Tracy, P., 125, 140*
Treynor, J. L., 79, 92*
Trice, H. M., 45, 47, 53*
Trist, E. L., 34, 53*
Tung, R. L., 44, 50, 51, 53*
Turoff, M., 127, 140*
Tushman, M. L., 9, 10, 15, 31*
Tyebjee, T. T., 75, 86, 92*

Utterback, J. M., 9, 15, 16, 29, 30*, 35, 53*, 95, 98, 117*

Vaill, P., 23, 31*
Van Wijk, G., 77, 91*
Van Maanen, J., 22, 31*
Van de Ven, A., 43, 53*
Velton, E. J., 59, 71*
Verdin, J. A., 127, 142*, 144–147, 149, 152, 156, 157, 158*, 159*
Vernon, M., 144, 145, 146, 148, 152, 155, 156, 159*

Von Thienen, V., 182, 183*

Wagner, G. R., 127, 142*
Wald, S., 99, 118*
Wallis, K., 106, 107, 119*
Walton, R. E., 189, 190, 196, 206*
Warner, K., 95, 119*
Waterman, R. H., 86, 92*
Waterman, R. H., Jr., 26, 31*
Weick, K. E., 42, 45, 52*, 53*, 54*
Weir, M., 127, 141*
Weirmar, K., 144, 159*
Werner, D. J., 137, 142*
Wernerfelt, B., 60, 72*
Wesa, L., 98, 113, 118*
Weston, J. F., 79, 92*
Whisler, T. L., 127, 142*
Wilde, E., 144, 145, 146, 148, 152, 155, 156, 159*
Wilkinson, I., 59, 72*
Williamson, O. E., 33, 35, 36, 40, 43, 44, 45, 54*
Wills, R., 95, 96, 99, 117*
Winter, S., 95, 118*
Withey, M., 195, 198, 199, 206*
Wolfe, B., 11, 14, 31*
Woodward, J., 189, 206*
Wright, R. W., 41, 54*
Wrigley, L., 79, 92*
Wysocki, B., 41, 54*

Yalowitz, M. S., 76, 91*
Yasai-Ardekani, M., 79, 90*

Zaltman, G., 56, 72*
Zmud, R. W., 127, 128, 142*
Zuboff, S., 128, 132, 142*, 144, 159*, 179, 185*
Zubrow, D., 163, 164, 184*

Subject Index

academic values, 9, 16–18, 21–24, 28
autocorrelation, 103–105, 107
automation, 188
autonomy, 187–190, 193–195, 198–200, 202, 205

balanced innovation strategy, 86
Box-Jenkins method, 103, 104
business cycle, 106, 107

Canada, 93–96, 111, 117
Canadian data, 98, 99
capital
  channel, 57
  goods industry, 77, 78, 82–84, 89
career
  opportunities, 143, 145, 146, 150, 151, 155, 156
  success perception, 161–163, 165–169, 172–174, 177, 178, 181, 182
CIM-computer integrated manufacturing, 192
coercion, 57–59, 61, 63–66, 68–70
commitment, 60 66, 67, 69
computer
  ease of use, 127, 128, 131, 132
  end user, 144–146, 148, 149, 153–158, 161, 175, 176, 180
  managerial use of, 123, 124, 126–132, 136–139
    level of, 123, 125, 136
    variety of, 136
  monitoring of users, 132, 143, 150, 155
  skill, 124
  use
    benefits from, 127, 138
    time savings from, 123, 127, 138
  computer-based technology, 161–163, 165, 168, 169, 174–177, 179–182
    intelligent workstation, 161, 164–166, 169, 174–177, 179–182
    main-frame terminal, 161, 163–166, 169, 174–177, 179–182
computer beliefs, 161–169, 174–182
  about communication, 161, 163, 166, 169, 172–180
    control, 161, 163, 166, 169, 172–180
    quality of job life, 162–164, 166, 169, 171–176, 178–180, 182
    work effectiveness, 161, 163, 166, 169, 171–176, 178–180, 182
consumer durable goods industry, 77, 78, 82–84, 89
consumer nondurable goods industry, 77, 78, 82, 83
contractual relationship, 36
Cooper Industries, 188
coping behavior, 190
core managerial work, 131, 139
covariance structure analysis, 130, 135
crisis of innovation, 74, 85
crosscorrelation, 103, 105–110, 114, 115
cross-cultural, 40, 45
cultural
  factors, 40, 49
  values, 40, 41, 46, 47, 50, 51
culture, 40, 47, 49, 50
  corporate, 41, 42, 45
  national, 41–43

dealing with others, 190, 193, 198, 202
decision making processes, 143–147, 149–153, 155, 157
defense contracting, 55, 56, 60–62, 64, 67, 69
demographics, 163, 165, 168, 176, 177, 179
deployment strategy, 191, 193, 195, 200, 202, 203, 205
detrending, 105–107, 110
diversification, 73, 74, 77–79, 81–84, 87, 88
dominant business diversification strategy, 79, 81, 82, 89
dominant design, 15–16

due process, 43, 44, 46–49, 51
 procedural, 43, 44, 46, 49–51
 substantive, 43, 44, 47–51

Eaton Corporation, 188
economic perspective, 33–35, 37, 38, 48, 49
electronic calendar, 131, 139
end-user computing, 144–146, 148, 149, 153–158, 161, 175, 176, 180
engineering perspective, 37, 38
EW – electronic workstation, 123–132, 136–139
external acquisitive growth strategy, 76, 78, 81, 83

factory of the future, 203
first-line, 187–193, 195–198, 200–205
FMS – flexible manufacturing system, 192, 196, 197
Ford-Rouge River plants, 193
foreman, 188, 189, 191, 192, 195, 196, 198, 200, 204
frequency domain, 104

GDP – gross domestic product, 100, 101, 106, 108, 109, 111, 114–116
gender, 161, 162, 164–167, 175–177, 179, 181, 182
global
 competitiveness, 74, 76, 77
 market (place), 39, 75, 88
governance, 46, 48, 49
 procedural, 46, 48
 substantive, 46–48
grand strategy, 78

hierarchy, 189, 200, 202, 205
high-performing systems, 23
high-tech research consortia, 29–30
human relations, 188
human resource management, 94, 96, 107, 112, 113, 116, 117

implementation, 125, 126, 130
improvisation, 14, 16
incremental product innovation, 15, 16, 25
inducement, 57–66, 68, 69
information technology, 123–126, 128, 132, 138, 139
innovation, 73–77, 85–88, 93–96, 98, 99, 103, 107, 111–113, 115, 116
 crisis of, 74, 85
 diffusion of, 55, 56, 60, 67, 70
 incremental product, 15, 16, 25
 measurement of, 95
 radical product, 9–11, 17, 21–24, 28–30
interdepartmental communications, 143, 145, 149, 150
internal growth strategy, 76, 78, 81, 83, 84, 89
interpretation systems, 33, 42, 44–46, 49, 51
investment, 60–62, 66, 67, 69

jazz format, 10, 14
joint ventures, 34, 37, 41, 44, 46, 49–51

labour productivity, 93–96, 109–111, 114–117
lag/lag structure, 104, 107, 108, 110, 111, 114
laissez-faire management, 20
liberating form, 9–11, 17, 21–24, 28–30

manipulation, 55–59, 62, 63, 67, 68, 70
manufacturing, 187–189, 191–194, 203–205
market returns, 73, 79, 80
measurement of innovation, 95
mechanistic structure, 36
mode, 57, 61
modernization, 188, 191, 195, 196, 200, 203, 204
multinational cooperative ventures, 33, 34, 39–43, 45, 46, 49–52
mundane first assignment, 18, 22, 27

negotiation, 33, 36, 42, 43, 45, 46, 49, 51
NIH – not-invented-here syndrome, 75, 85, 86, 88

obligation, 57–59, 61, 63–66, 68, 69
occupations, 195–198, 201, 202
office automation, 124, 126
open design, 9, 14, 16, 17, 21, 25

organization(al)
  context, 146
  structure(s), 34, 36, 204, 205

Parsons, John, 194
perception, 55, 59, 60, 64–66
  of career success, 161–163, 165–169, 172–174, 177, 178, 181, 182
    financial, 162, 166, 169, 170, 173, 178, 181
    general, 161, 162, 166, 169, 170, 172–174, 178, 181, 182
    interpersonal, 162, 166, 169, 170, 173, 178, 181
    job-related, 161, 162, 166, 169, 172–174, 178, 181, 182
persuasion, 56–59, 61, 63–66, 68, 69
principal industry, 79, 88
producer goods industry, 78, 82–84, 89
product life cycle, 15, 25
productivity, 93, 94, 98, 101, 102, 107, 111–113, 117, 143, 144, 146, 147, 150, 152–158
professional
  managers, 19, 24
  perspective of R & D administration, 34, 35, 38, 49
  values, 17, 22

quality of output, 143, 144, 146, 147, 149, 150, 152–158

radical product innovation, 9–11, 14–17, 21–23, 25, 26, 28–30
R & D – research and development, 33–42, 44–46, 48, 49, 51, 52, 73–77, 80–89, 93–102, 106–116
R & D intensity, 73, 81, 82, 84, 89
related business diversification strategy, 79, 81, 82
retrenchment strategy, 78, 81, 83
role stress, 188–190, 193, 195, 196, 198–203, 205

selection, 22, 27
self-selection, 18, 22, 27
single business diversification strategy, 79, 81, 82
skill requirements, 143–146, 149, 150, 155, 156
socialization, 20, 22, 24
sonnet form, 11
stability growth strategy, 78, 81, 83, 84, 89
Star Electronics, 17–25, 27–29
strategic
  alliances, 75
  posture, 74
strategy, 55–71
  balanced innovation, 86
  deployment, 191, 193, 195, 200, 202, 203, 205
  dominant business diversification, 79, 81, 82, 89
  external acquisitive growth, 76, 78, 81, 83
  grand, 78
  internal growth, 76, 78, 81, 83, 84, 89
  related business diversification, 79, 81, 82
  retrenchment, 78, 81, 83
  single business diversification, 79, 81, 82
  stability growth, 78, 81, 83, 84, 89
  unrelated business diversification, 79, 81–83
stress, role, 188–190, 193, 195, 196, 198–203, 205
structure(s), 22, 34, 36
  organization(al), 34, 36
supervisor, 187–193, 195–198, 200–205

task, 35, 47
  complex, 35
  ambiguous, 35, 47
teams, cross-functional, 204, 205
technical guru, 23, 24
technology, 187–193, 195, 197, 199, 200, 202–05
  transfer, 28–30, 113
  workgroup, 193, 195, 198–202
technological change, 93–95, 97, 98, 101, 111
time domain, 103, 104
time-series analysis, 100, 103–107, 110
Timken Company, 188

unarticulated organizational goals, 19
uncertainty, 34, 35, 40

university/corporate research, 28, 30
unrelated business diversification strategy, 79, 81–83

values, 9, 16–18, 21–24, 28, 40, 41, 46, 47, 50, 51
  academic, 9, 16–18, 21–24, 28
  cultural, 40, 41, 46, 47, 50, 51
  professional, 17, 22
ventures 33, 34, 41, 45–48, 50–52
  joint, 34, 37, 41, 44, 46, 49–51
  multinational cooperative, 33, 34, 39–43, 45, 46, 49–52
  venturing agreements, 37
vision of ultimate technology, 25

weapons systems, 55, 56, 59, 67–69

de Gruyter Studies in Organization

An international series by internationally known authors presenting current research in organization

Vol. 1

The Japanese Industrial System

By *Charles J. McMillan*
2nd revised edition
1985. 15.5 x 23 cm. XII, 356 pages. Cloth. ISBN 3 11 010410 5; 0-89925-005-X (U.S.)

Vol. 2

Political Management

Redefining the Public Sphere
By *H. T. Wilson*
1984. 15.5 x 23 cm. X, 316 pages. Cloth. ISBN 3 11 009902 0

Vol. 3

Limits to Bureaucratic Growth

By *Marshall W. Meyer* in Association with *William Stevenson* and *Stephen Webster*
1985. 15.5 x 23 cm. X, 228 pages. Cloth. ISBN 3 11 009865 2; 0-89925-003-3 (U.S.)

Vol. 4

Guidance, Control and Evaluation in the Public Sector

Edited by *F. X. Kaufmann, G. Majone, V. Ostrom*
1985. 17 x 24 cm. XIV, 830 pages. Cloth. ISBN 3 11 009707 9; 0-89925-020-3 (U.S.)

Vol. 5

International Business in the Middle East

Edited by *Erdener Kaynak*
1986. 15.5 x 23 cm. XVI, 278 pages. Cloth. ISBN 3 11 010321 4; 0-89925-021-1 (U.S.)

Vol. 6

The American Samurai

Blending American and Japanese Managerial Practice
By *Jon P. Alston*
1986. 15.5 x 23 cm. XII, 368 pages. Cloth. ISBN 3 11 010619 1; 0-89925-063-7 (U.S.)

Vol. 7

Organizing Industrial Development

Edited by *Rolf Wolff*
1986. 15.5 x 23 cm. XI, 391 pages. Cloth. ISBN 3 11 010669 8; 0-89925-168-4 (U.S.)